THE HANDBOOK OF
BUSINESS
SECURITY

THE HANDBOOK OF
BUSINESS
SECURITY

A PRACTICAL GUIDE TO MANAGING THE SECURITY RISK

KEITH HEARNDEN & ALEC MOORE

Second edition

IOD

INSTITUTE OF DIRECTORS

KOGAN PAGE

ACKNOWLEDGEMENTS

The authors would like to thank Ian Dobson for his invaluable contribution on transport security, and also those who checked the accuracy of the passages dealing with security technology.

The publication you are reading is protected by copyright law. This means that the publisher could take you and your employer to court and claim heavy legal damages if you make unauthorised photocopies from these pages. Photocopying copyright material without permission is no different from stealing a magazine from a newsagent, only it doesn't seem like theft.

The Copyright Licensing Agency (CLA) is an organisation which issues licences to bring photocopying within the law. It has designed licensing services to cover all kinds of special needs in business, education and government.

If you take photocopies from books, magazines and periodicals at work your employer should be licensed with the CLA. Make sure you are protected by a photocopying licence.

The Copyright Licensing Agency Limited, 90 Tottenham Court Road, London, W1P 0LP. Tel: 0171 436 5931. Fax: 0171 436 3986.

First published in 1996
Second edition 1999

Kogan Page Limited
120 Pentonville Road
London N1 9JN
UK

Kogan Page Limited
163 Central Avenue, Suite 4
Dover, NH 03820
USA

British Library Cataloguing in Publication Data
A CIP record for this book is available from the British Library.
ISBN 0 7494 2923 2

Typeset by Saxon Graphics Ltd, Derby
Printed by Bell & Bain Ltd, Glasgow

CONTENTS

WHY BOTHER WITH SECURITY?

It is a perfectly legitimate question to ask why we should bother with security. In the process of answering it, we shall indicate some of the aims this book sets out to achieve, as well as others that form no part of our aims. The straightforward answer to our question 'Why bother with security?' is 'Because it can help you to improve the bottom line'.

The purpose of this book is to bring to your attention some of the important ways in which *in*security can threaten and undermine the performance of your business and to suggest common-sense, practical ways in which you can combat these threats. Above all, the authors recognise that you have a business to run, that the time you have available to consider and react to the kinds of events we shall describe is limited, and that the resources you can make available to counter the threats are strictly finite. We also recognise that you are not likely to be a victim of all the criminal methods we shall describe in the book; so turn to those chapters that appear to offer you the prospect of the greatest return on your investment – which, most importantly, is your time.

We do not set out to argue that the world in which we all live and operate is so dangerous, so full of crime and criminals, that, as a manager responsible for business success, you must buy heavily into security as the only means to counter such threats. We believe that an awareness of *in*security within the business helps you to identify and measure both the extent and the likelihood of the threats you may face, following which you can make a more informed judgement about the impact they might make on your business, should they materialise. To rephrase this in terms favoured by current use, **we intend this book to help you manage risk.** 'Managing risk' is a vital and constant element of managing a business. Although it commonly includes finance, investment, trading in general, insurance and safety, risk management should also embrace the *security* aspects of risk. If not, there is a likelihood that business performance will not be as good as it might – and should – be.

CRIMES AGAINST BUSINESS

Before moving on to the main substance of the book, let us try to put the problem of business crime (that is, crime directed *against* business) into context. During 1994 the following crimes with a business or public service relevance were recorded by the police (Home Office, 1995):

Non-domestic burglaries	578,300
Robberies	59,800
Thefts from shops	269,300
Theft by an employee	13,786
Other theft or unauthorised taking	554,800
Fraud and forgery	145,800
Criminal damage (over £20 value)	677,500
Arson (non-domestic)	30,600

Although these are horrendously high figures, they only represent those crimes actually *reported* to the police. Evidence produced in the 1992 British Crime Survey (Home Office, 1993, pp 11–24) indicated that many crimes were never reported to the police, so that the official statistics are an understatement of the extent of crime. In the important retail sector, for example, the 1993–94 survey by the British Retail Consortium (Speed, Burrows and Bamfield, 1995, p 33) revealed that only 68 per cent of customers arrested for theft from shops were passed to the police for processing, whilst even fewer staff thieves (46 per cent) were reported.

Within the world of business, personal experience leads one to believe that very many crimes committed by employees – especially 'white collar' crimes like fraud, embezzlement and theft of information – are never brought to the notice of the police. It is easy to understand the reasons for this, even though one might disagree with them. Businessmen are quick to appreciate the possible loss of public confidence, for example, if they were to reveal that one of their employees had stolen a significant sum of money from them. Often, therefore, seeking restitution of the amount stolen is a priority, followed by a quiet dismissal of the offender. This course of action also avoids the disruption that can occur from any police investigation that might follow a reported suspicion. The problem with this approach is that there is little deterrent to others who might also be tempted to steal from the organisation, because the penalty of being found out is perceived to be small.

What is certain, however, is that by any measure one cares to adopt, there is an extremely serious problem for businesses to confront and overcome. In the remainder of this chapter, we shall consider various aspects of business crime, and by so doing we hope to set the scene for what follows in the rest of this book.

CASE STUDY 1

Imagine that you are a sole trader running a shop that sells sports goods and clothing. Your sales margins are 60 per cent, your annual sales are £200,000 and your business overheads are £75,000. You should, therefore, be making a profit of £45,000; but when your accountant has completed his annual audit, he tells you that you only have £30,000 to show for your year's labours.

So where has the missing £15,000 gone? Have shoplifters taken it (an obvious first thought)? Have John or Jane, your two shop assistants, stolen it (a reluctant second thought)? Were there really that many dud cheques taken during the year (an immediate third thought)? Have the suppliers or, more likely, the delivery drivers, not actually delivered the number of cricket sweaters/badminton rackets/tennis balls/Manchester United shirts shown on the advice notes and invoices (a possible fourth thought)? Or has Sylvia, your part-time bookkeeper, somehow managed to 'fiddle the books' (the only thought left)?

The bottom line is that you have failed to acquire £15,000 that should rightfully have been yours to live on, treat yourself, invest, or to dispose of as you thought fit.

CASE STUDY 2

Now imagine that you are the General Manager of a medium-sized electronic components manufacturer – you have a targeted annual turnover of £100 million, with a budgeted profit of £12.5 million before Head Office charges. Yours is a responsible job, for which you are paid a basic £65,000 a year, plus a performance-linked bonus based upon 1.5 per cent of every additional £100,000 profit achieved over the target figure.

Your personal expectation and aim at the start of the year are to exceed your target turnover by 10 per cent, thereby improving your overall margins to 13 per cent, resulting in a trading profit of £14.3 million. This is £1.8 million over your target, thereby achieving for you a personal bonus of £27,000.

At the end of the year you have increased sales by only 9 per cent, but you have more than compensated for this shortfall on your personal target by managing to reduce employee numbers by 65, saving £1.7 million in this way. Of course, from time to time, you have had to hire in some contract agency staff to see you through the main holiday periods, but overall you are satisfied that you are ahead in the numbers game that Head Office insists on nowadays.

It comes as an enormous shock and disappointment, therefore, when your company's audited profits are stated to be, not the £14.3 million you were expecting, but only £11.8 million. Even worse, this is below your official target set by Head Office; so, far from looking forward to a bonus, your anticipation now takes on a much less pleasurable prospect.

How could you possibly be £2.5 million worse off than expected? Where should you look? You are, of course, aware of reports suggesting that business crime is a growing problem, and in particular you remember reading about high-tech crime that depends upon the misuse of computers and computer networks, sometimes associated with computer hacking. Then there was the court case involving the theft of sensitive information from leading oil companies, which somehow was exploited to undermine the tendering process for the construction of North Sea oil rigs. Now you come to think about it, the reason for your market edge lies mainly in the innovative quality of your manufacturing processes – and knowledge of that would be worth something to your competitors! Remembering how you had to hire in two production engineers to cover for sickness and holidays, you begin to worry whether the agency did bother to check references before sending them. What's more, you operate in a limited market, where people with the required skills are hard to find, so you now wonder where those two engineers went to work once they left you.

And then there is fraud, of course. There have been quite a few reports by leading firms of management accountants that indicated how frequently businesses were becoming the victims of fraud and, what is more, expected the problem to get worse over the next few years. Theft of company property, you now acknowledge, might well be occurring in your business, as well as in the food processing company on the other side of town, which brought a case (reported in the local paper) against 20 of their employees for stealing packaged meals to order and selling them in the local pubs. You now realise that there might well be a similar market for your own company's electrical and electronic components.

Still searching for reasons for the missing £2.5 million, you begin to wonder whether you have been getting full value for the scrap by-products of your engineering processes. Who has been given responsibility for this? At the other end of the cycle, what systems and checks are in place to ensure that your buyer is obtaining the keenest prices for your raw materials? He must be responsible for spending upwards of £20 million a year, which is more than enough to make him a target for some of those unscrupulous suppliers out there.

The more you ponder on the missing profit, the more you begin to realise how many ways that money could have disappeared. Not all at once, of course, for that would have been noticed quite quickly; but trickling out here and there, and with insidious effect on the one thing above all that shareholders, directors and managers live by – the bottom line.

THE EFFECT OF DOWN(RIGHT)-SIZING ON EMPLOYEE CRIME

A combination in the early 1990s of the down side of the international economic cycle and the first full flowering of 'the information revolution' caused by massive advances in computing and communications technology, has resulted in a widespread restructuring of business organisation. The internationally depressed trading situation also coincided with a rise in consumer expectations about the quality of the products it was expected to buy, thereby increasing the pressure on businesses to provide more for less, and causing boards of management to intensify their search for operating economies.

Meanwhile, the universal adoption of the Personal Computer (PC) meant that business information could be acquired, collated, analysed and recompiled with a speed and facility never known before. Add to this the ability to communicate that information electronically, and therefore at great speed, to other managers within the organisation, to trading partners in joint ventures and to customers, and it can be easily perceived how the process of managerial decision-making was being fundamentally reshaped. What became clear, as these changes were embedded in the way business was conducted, was that the computer had taken over the task of handling information that was previously the responsibility of several layers of middle-grade managers. Senior managers came to realise that there was no longer a need to employ a whole group of managers whose principal task, to put it crudely, was to receive information, process it in some predefined way, and then to present it to the management level above them. There was no longer a need to employ a whole group of managers, because senior managers could arrange for the information to be sent to their computer direct and then, with appropriate software, all the processing previously carried out for them at lower levels could be conducted within the organisation.

These senior managers remained, by and large, because it was they who were making the business decisions. However, at the management levels below them there has been a great shake-out. The main difference this time, compared to those periods in the past when businesses were forced to make large numbers of employees redundant, is that it is the well-educated, professional and middle-grade managers who have been the principal casualties. This is an intelligent, astute and vociferous group, who have by and large correctly rationalised their position as being that of blameless victims of an inexorable process. They have also identified a marked discrepancy in the treatment they receive, compared to that awarded to the most senior executives in their companies, whose three- and five-year rolling contracts are used to ensure a handsome severance package, whenever their association with the employing organisation is ended. This even occurs, they will argue, when it is precisely those senior executives whose poor decisions and leadership have produced the conditions that made their own redundancies necessary.

So widespread have been managerial redundancies that few of those in work regard their own position as inviolable, most have experienced the loss of former colleagues and a great many have come to recognise that a long-term, career-enhancing relationship with a single employer is not an option for the future. These are conditions that can readily create stress among the managerial group. Such conditions are also likely to break down those traditional bonds of loyalty that developed between employer and employee as a consequence of a perceived trusting and long-term relationship, based upon mutual interest and support.

Accompanying this now well-established trend to reduce the numbers of employees, there has also been a growth in the use of short-term contractors. The one action tends to be consequent upon the other; with senior management from time to time having to acknowledge that too few key personnel remain to complete the corporative task, and therefore seeking to make good the shortfall by hiring temporary help.

Such action is fraught with insecurity, as appears to have been the case with a much publicised event at British Telecom. Here, the temporary worker allegedly made no secret of the fact that he was a freelance journalist, but was nevertheless hired by BT. In order to enable him to work productively, he appears to have been given privileged access codes to BT's computer database of sensitive and secure telephone numbers, including those of the Prime Minister and Government intelligence agencies. Not altogether surprisingly, he gained extensive media publicity for his claim that BT's most secret telephone numbers were kept insecurely and had been *hacked* and down-loaded on to a computer bulletin board. At the time of writing, the case is believed to be *sub judice,* but the resultant publicity prompted questions in Parliament and much adverse media speculation about the quality of BT's security.

There is thus a growing fear, initially voiced by security professionals, but now beginning to be taken up more widely, that the increasing casualisation of employment creates conditions that are likely to encourage criminal activity, as some managers react against the perceived injustice and insecurity of their situation and others are replaced by temporary contractors. If this view proves correct, the security implications are very serious, for the threat derives from a group of people who are strong on technical skills and provided with ample opportunity, but who are at the same time only loosely supervised. Fraud, theft of sensitive corporate information, purchasing scams, sabotage of the corporate computing system and other 'white collar' crimes are seen to be the areas of prime risk from the disillusioned manager.

It is possible, too, that the search for operating efficiencies through constantly reviewing the exact nature of the contribution achieved by specific posts and post-holders within the organisation may lead to the abandonment of some important business controls. This may be done wittingly or unwittingly. A prime example of what can and does happen is often evident in the way in which certain business operations are transferred to computer processing. In accounting, for instance, a fundamental control is one that calls for the separation of the task of handling goods or monies from that of recording the transaction. It is easy to see what might happen if the same employee were given the authority to sanction the payment of invoices received for the supply of goods or services and also the right to initiate and authorise the cheques raised as a consequence. An obvious control, you might think, but the annals of computer-linked fraud contain innumerable examples where this fundamental control was missing. It is certain that its abandonment was caused in many cases because someone said, *'Why do we need two people to run our purchase ledger, when one could do it perfectly well now that we have a computer to speed things along?'*

It is not the purpose of this book to challenge the decisions of countless boards of management to meet the intensity of competition, at least in part, by reducing the numbers of their employees. However, it is legitimate to present the argument that there may sometimes be a price to pay for this apparent gain in efficiency, and that price is an increase in the organisation's exposure to criminal attack through a decline in security.

ON BEING A VICTIM

Perhaps you have already been a victim of a crime. In the light of the very high numbers of crimes committed annually against the business community, that would not be surprising. However, it is at this point worth reflecting on some of the consequences of being on the receiving end of criminal activity, if only to reinforce how much pleasanter it is to forgo the experience.

The most obvious consequence is that there is almost inevitably a financial cost. As we noted in the two scenarios described at the beginning of this chapter, undetected crime ultimately shows up in the balance sheet and in the profit and loss account (with the emphasis on the *loss*). How long it remains undetected is likely to affect the severity of the shock, but is unlikely to avoid the experience of shock itself.

Perhaps a personal experience from many years ago, when I was working as an Organisation and Methods Manager in Yorkshire, will illustrate the typical reaction. I had been with the firm, a multiple dry-cleaners with a substantial factory and head office complex, for only a few months, when the Managing Director asked me to look into the reasons for an over-budget subsidy to support the works canteen. What I eventually found, but could not at first believe, was a systematic fraud being perpetrated by the cashier in the Accounts Department. Bills for food delivered by local tradesmen to the canteen were brought to him and were paid in cash there and then. Most of these bills were handwritten. So a delivery of potatoes, costing £7.50 (it *was* a long time ago!) was paid from Petty Cash and the receipted invoice was used to support the payment. What the cashier did was simple, but effective. It was simple in that all he did was to add a figure '1' before the '7', making it appear as though the potatoes cost £17.50, and pocket the £10 himself. Everything balanced. It was effective, because no one until then had tested for the *reasonableness* of the cost of potatoes, nor had anyone directly supervised the cashier's work.

My immediate reaction was to think that there had been some kind of mistake, for which there was a perfectly logical explanation. After all, I knew the cashier; we sometimes had lunch together. It couldn't possibly be a deliberate crime. But crime it was, and he confessed all to the police. About £1500 had been stolen in this way during the previous year. It was not a huge sum. It did not threaten the viability of the business, but it had caused the MD to consider putting up the price of the canteen meals, which might well have had an adverse effect on the morale and attitude of the factory employees. The discovery of the crime also had wider repercussions (which I am reporting, not justifying). The MD held the cashier's immediate superior, the Financial Accountant, culpable for not discovering personally what had been going on, and he was dismissed. The same fate also befell the Chief Accountant, who was on holiday at the time, because ultimately, the MD argued, it was down to him to establish the controls and systems within the department so that frauds of that kind became impossible.

So you can see from this simple example just how much personal angst and organisational trauma directly resulted. There was, too, considerable disruption to the business, in that it was suddenly shorn of its top three accountants, leaving a sadder and a wiser O&M Manager in temporary charge until replacements could be recruited – a period that eventually lasted for nine months. Of course, not all organisations or Managing Directors will react in such an extreme way, but there is still a marked tendency for people to believe that *'It won't happen to me'*, when all the evidence suggests otherwise. I did not at the time attempt to calculate the cost of all this; but looking back, there was little doubt that productivity in the Accounts Department suffered for a while: there was the severance pay to the Financial and Chief Accountants, the cost of recruiting their successors, the cost of senior management time, and the cost of paying me to do a job for which I was not well qualified, whilst I was unable to do the job I was good at.

Although it is extremely difficult to implement security measures that prevent all theft and fraud, there is a strong argument to suggest that in this case a basic awareness of the possibility of fraud, accompanied by a segregation of the task of recording the cash payments from that of handing over the actual money would have made the crime very difficult to sustain. Furthermore, if a crime occurs despite such basic safeguards, good security would devise a way of signalling that something was amiss in sufficient time to minimise the loss and disruption. Systems to achieve this are well established and available, and will be discussed in greater detail in Chapter 6.

ON SUFFERING A DISASTER

Sometimes events occur that have a disastrous effect upon the organisation, causing severe, often prolonged disruption to its operations. As a consequence, one of the key tasks nowadays in any enterprise is business continuity planning, a role that nearly always has security implications.

Disasters come in many forms, but their possible occurrence can usually be identified in advance, allowing contingency plans to be developed to cope with any emergency and to minimise its effect. Some reasonably common causes of major disruption are fires, prolonged interruption of computing facilities, floods, terrorist acts, deliberate sabotage of plant, equipment and premises, a lengthy labour dispute and the loss of key personnel. To this list, one might add the impact of a really major fraud or theft of company assets, such as was experienced by BCCI and the Maxwell Corporation.

There is a tendency to believe that you can insure against such catastrophic events, which to some extent is true. The problem is that insurance is expensive to arrange in the first instance, carrying with it the high probability that the insurer will insist on the implementation of a range of countermeasures to minimise his potential exposure. Secondly, the compensation ultimately provided by insurance is very unlikely to be available immediately after the catastrophic event and is therefore of little help in getting the business up and running again – paying the staff and the suppliers, or renting new premises and equipment. At worst, by the time recompense has been received, there may be no business left.

One of the under-reported tragedies of the bombing of the World Trade Center in New York was the subsequent failure of numerous small businesses operating within the vast complex, whose chosen tasks were to perform a variety of services to the employees of the larger organisations with offices there. These small concerns ranged from the individual 'shoe-shine boy', to dry-cleaners, delicatessens, cafés and restaurants, quick-print shops, and so on. The problem for these service providers derived from the prolonged closure of the building whilst the extent of the bomb damage was surveyed and repaired. This denied them access both to their premises and, more critically, to their customers. Their means of trade, therefore, was removed at a stroke; they were denied their sole source of income, and they failed.

So contingency planning for disaster recovery is a key aspect of risk management, whilst managing security risks is an essential subset. All managers should at some stage find time to ask a series of 'What if?...' questions, in which they explore the consequences of a variety of scenarios. There are, as usual, a number of tools to help them, and these will be considered later. However, it is only by asking the questions that potential key exposures will be identified, enabling plans to be developed to contain any emergencies and organise an effective response.

THE DUTY OF CARE

Quite apart from any responsibilities that fall on the directors of limited companies to safeguard corporate assets, there is also a specific legal responsibility under the Health and Safety at Work Act (HASAW) that managers take all reasonable measures to ensure the safety of employees. This clear provision often has security implications, for there are a number of areas where it is a criminal or quasi-criminal act that threatens the physical or psychological health of the employee.

Recent years have witnessed a worrying increase in the numbers of incidents of violence towards staff. Those in occupations intended to provide a service to the public have been particularly at risk. Many jobs in the public sector fall into this category, such as Local Authority Housing Benefit Officers, Social Security Officers, public transport staff, nurses and paramedics (especially those in Accident and Emergency Departments), social workers and schoolteachers. For some time, though, the problem has also extended to shop and store employees, bar and hotel staff, staff running recreational and leisure facilities, and those working in retail financial services, such as banks and building societies. To provide some idea of the scale, retailers reported that 12,055 staff were subjected to physical violence in 1993–94, whilst another 90,421 were subjected to threats of violence and 209,645 were the victims of verbal abuse (Speed, Burrows and Bamfield, 1995, p21).

Managers who are anxious to preserve the safety of their staff from the threats posed by the public are reacting to a security problem. Their solutions might include physical security (providing barriers), access controls, monitoring (through CCTV surveillance), panic buttons, security guards (visit your local hospital's Accident Department on Friday or Saturday evening to witness this approach), mobile communications (radios or telephones to ensure speed and ease of contact with a support service), and changed systems of working.

In an increasing number of occupational areas, therefore, thinking about security can positively enhance staff safety and, by logical extension, improve their morale and productivity. Although responding to the threat of violence can be seen as a strong moral imperative, it is also likely to achieve benefits of a more practical kind.

SUMMARY

What we have tried to show in this chapter is that there are a number of important ways in which questions of security are inherently present in the successful management of a business. The levels of crime and the variety of the threats to both businesses and the public sector are such that those with the responsibility for managing them must be in a position to respond effectively if they are to realise the full potential of the people and the organisations they lead. The successful use of security techniques and technology will usually result in the reduction of business or operational losses, which in turn will convert into an improvement of the final profit or a more effective use of scarce resources.

However, relatively few companies have ready access to a source of security expertise, so the principal benefit we hope you will derive from this book is an enhanced ability to observe your business from a security perspective; you will become more aware of the various risks to the business, understand more completely the criminal methods likely to be employed against you, and be equipped to respond positively and effectively to such threats.

References

'Notifiable Offences England and Wales, 1994', Home Office Statistical Bulletin Issue 5/95, Home Office, London, 11 April 1995.
'The 1992 British Crime Survey', Home Office Research Study 132, HMSO, London, September 1993.
Speed, M, Burrows, J and Bamfield, J (1995) 'Retail Crime Costs: 1993–94 Survey', British Retail Consortium, London.

2

THE SECURITY SURVEY

Managing the Risks

INTRODUCTION

In Chapter 1 we discussed why security is necessary and indeed why we ignore it at our peril. The question is, *'How do we go about ascertaining how much, or how little, we need and where should it be applied?'* The technique is known as **the security survey.** A full survey will consist of three distinct phases. First, we must *identify* the assets and profit-making activities within the organisation which are exposed to risk. Next, we must *evaluate* how serious the risks are, and finally, we must prepare and implement a system to *manage* those risks. Each of these three elements will be examined shortly; but first, let us define what we mean by *risk*.

In the *Kluwer Handbook of Risk Management* (Croner Publications, Supp. 6: 1.1–09) a clear distinction is drawn between *business* and *pure risks*, although there may be an overlap in some places:

'Damage to a company's assets through purely economic losses such as the general level of business activity, alteration in foreign exchange rates, inflation, technological progress, etc falls outside the scope of risk management.'

Pure risk, on the other hand, is defined as:

'the preservation of the organisation's assets and earning power from sudden losses'.

It is in the context of pure risk that we will now examine the concepts of risk identification, evaluation and management.

RISK IDENTIFICATION

Figure 2.1 illustrates a simple but effective approach to risk identification *(Kluwer 4.1–03).*

CAUSE		EFFECT

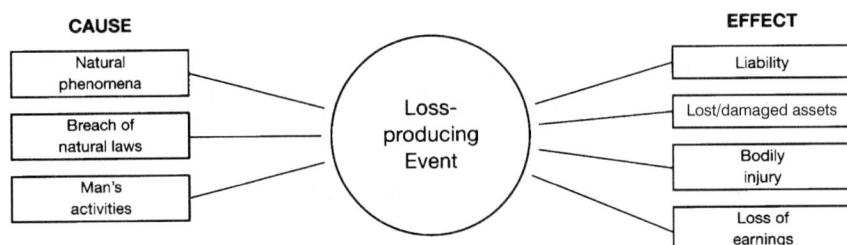

Figure 2.1 *A simple but effective approach to risk identification*

The figure indicates three steps in the process of identification:

1. Perception – the identification of a possible loss-producing event.
2. The analysis of possible operative perils and impinging hazards.
3. The consequential loss effects.

In order not to overlook any major sources of risk, we need to examine the prime areas of the business where most losses occur:

Physical assets. Such assets will include property, machinery, products, raw materials, cash, cheques etc and will be at risk to theft, damage or destruction, or possibly, in the case of some types of products, contamination.

Personnel. Some organisations are heavily dependent upon key members of staff. Others unavoidably place staff in positions of risk – for example, bank counter staff, sales representatives travelling into volatile countries or rent collectors. The loss, injury, sickness or death of staff may have a profound effect on the health and wealth of a company.

Information. Frequently referred to as the life-blood of a company, the importance of the integrity, availability and confidentiality of proprietary information is often underestimated. Increasingly, such information is held in computers and the main dangers and weaknesses are examined in detail in Chapter 8.

Liability. Companies may be held legally liable for incidents involving injury to third parties (including employees) or damage to their property.

Business interruption. Such losses would generally flow from property, liability and personnel losses, where quite frequently the true cost to the company through loss of sales or production may far outweigh the material damage or liability loss. Contingency planning for business continuity is considered in Chapter 10.

Goods in transit. This extremely vulnerable area is discussed in Chapter 9.

Each of these areas needs to be examined critically and thoroughly if we are to produce an accurate and complete picture of possible loss-producing events. The person undertaking the security survey should have a sound and comprehensive knowledge of the nature of the business and be familiar with the extent to which the organisation relies on key services such as public utilities, as well as in-company processes and operations. One effective approach is to ask the relevant manager or director to identify what event or events would most seriously jeopardise the achievement of business objectives, for the answers will identify the key risks the organisation faces. The next task is to list these in an order which reflects their importance and priority, but to achieve this we need to **evaluate** each risk so identified.

RISK EVALUATION

Risk can be defined usefully as the product of:

1. The **probability** (or likelihood) of an unwanted event occurring.
2. The **cost** to the organisation or company if it does occur.

One can often find a link between the two – for example, low-severity events may occur more frequently than high-severity events. However, it may be that the low-frequency but high-severity events represent the most serious threat to the company's well-being. On the other hand, low-frequency events are often less predictable and thus more difficult to forecast. It quickly becomes apparent, therefore, that the first problem is one of gathering information about the *likelihood and impact of uncertain future events.*

Although various statistical methods are available for predicting the relative probability and impact of future unwanted events (discussed later in this chapter), the most useful is often the collection of data on **past** events. In-house and personal experiences, local crime statistics and background information, provided they are collected and collated, form a valuable basis for judgements about future events.

This is an appropriate point at which to examine an apparent contradiction, represented by the difference in the crime figures produced by the police and those published following the biennial *British Crime*

Survey (BCS, 1998). Every year a compilation of police records of notifiable offences is published (*Criminal Statistics*). Unfortunately, the police do not record all reported crimes. This may be because they are unconvinced of the authenticity of some reports, or perhaps they consider them too trivial to record. However, it is likely to be the victims' failure to report their experiences to the police that is the principal cause. The reasons for this reluctance vary from the relatively small impact of a crime on the victim to a general feeling that the police will be ineffective in dealing with the crime. Nor is it just individuals who are reluctant to report crime: companies will often deal with internal criminal activities in-house, rather than risk exacerbating the problem by courting adverse publicity. It may even be that a manager who has personally contributed to the crime through incompetence or lack of controls will assist in the concealment process in order to safeguard his or her own future within the company.

Every two years, the Home Office conducts a survey by interviewing adult members of the public (almost 15,000 last time) about their personal experiences of crime during the previous year, and the results are published in the *British Crime Survey*. This tends to provide a more accurate picture of crime in general. Even so, it does not reveal the total picture, since the person interviewed may not wish to discuss a past incident, particularly if this was not reported to the police. Nevertheless, the *BCS* statistics indicate a crime rate up to three times greater than that indicated by the official statistics of recorded offences.

Crime prevention panels, local authority records and local trade and industry organisations are other fruitful areas to explore when gathering information on crime statistics. One should look for emerging patterns, such as the times and places that are indicated as the most likely for specific criminal events, the types of offences committed and their *modus operandi*, any specific circumstances that apply, such as the influence of alcohol or drugs, or a particular profile of offenders and victims.

A simple but effective technique of examining possible **sources of threats** is to divide them into two groups – external and internal. *External* threat sources will include professional villains, special or single cause protest groups, terrorist organisations, opportunist thieves and vandals. *Internal* threats will come mainly from employees, but one should not ignore contractors or visitors. In his book on corporate fraud (Comer, 1985) Comer suggests that fraud (or theft) results from a combination of motivational and situational factors in which the critical point is the presence of an opportunity. Such an opportunity, he contends, will only be exploited by a criminal when he perceives a low chance of detection. The connections between fraud/theft and opportunity/motivation are shown in Figure 2.2.

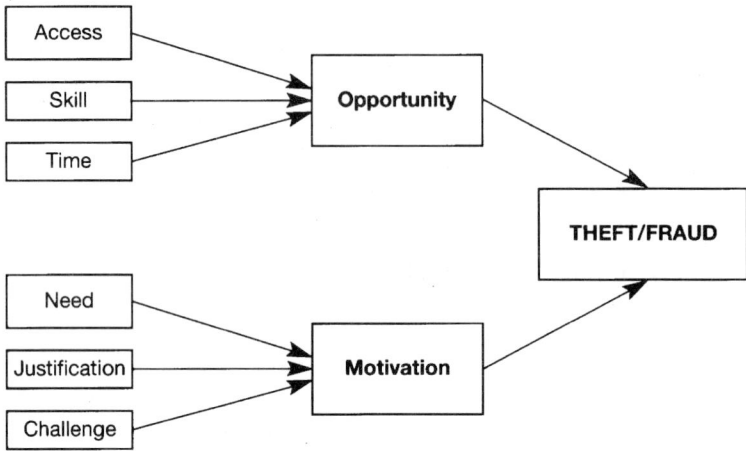

Figure 2.2 *Pre-conditions for theft*

The likelihood of an unwanted event occurring will therefore be directly influenced by the culture of the company. If there is ample evidence of tight control and clearly established procedures, the probability of criminal activity will be proportionately lower than within an organisation where chaos reigns.

The next step in risk evaluation is to determine the cost to the company when the event occurs. The severity of the event will depend on the value exposed and the degree of damage suffered. The value exposed has several components (Bannister and Bawcutt, 1982, p 49):

- The value of the property at risk.
- Any additional expenses that would or could result from the event.
- Loss of earnings due to non-availability of productive assets.
- Any damage or injury to third parties that could result in further financial loss.

It is clear, then, that *the loss may exceed 100 per cent of the nominal value exposed.* This is normally referred to as consequential loss.

Having identified the possible loss-producing events, examined the causes, predicted the probability and estimated the effect (cost), we can now categorise these risks in order of magnitude. Table 2.1 shows a simplified table of pure risks.

Table 2.1 *Methodology for ranking pure risks*

Event	Frequency	Potential cost	Cost per annum	Impact Severity Rating
1	Weekly	£50	£2,600	2
2	Monthly	£100	£1,200	4
3	Annually	£1,000	£1,000	5
4	Every 5 years	£10,000	£2,000	3
5	Every 10 years	£100,000	£10,000	1
		Total	**£16,800**	

Such a table will assist in allocating resources appropriately. It will also enable us to determine a cost/benefit ratio, for if we find that we are spending more than £16,800 per annum controlling the above risks, perhaps we should be re-examining our risk management strategy.

A risk exposure calculated along these lines may contain figures which have been arrived at largely by a mixture of intuitive and subjective processes. More reliable results might be achieved by using a statistical and quantitative method of risk analysis, such as:

CRAMM. CCTA* Risk Assessment and Management Methodology.

MARION. Risk assessment methodology created by and for the French insurance industry

Dependency Modelling Tool. A PC-based software package which, by showing the consequences of alternative actions or structure and their interdependency, makes it possible to engineer the most robust solution to an identified risk.

There are two distinct advantages in using such systems to analyse risks:

Funding for the loss reduction programme is more likely to be approved if the case is supported by a well-presented set of statistical data.

With the advent of computers and appropriate software programmes, such statistics are more easily compiled and presented. It is also a simple matter to keep the data up to date to cater for changing conditions and risk factors.

This brings us neatly on to the third and final phase of our survey.

* CCTA = Central Computing and Telecommunications Agency (of HM Government)

RISK MANAGEMENT

There are four common and interdependent elements that guide risk management – avoidance, transfer, retention and reduction.

Risk Avoidance

At first glance, risk avoidance may appear to be the answer to most of our problems. You *can* avoid the risk completely, for instance, by ceasing to engage in that particular activity which gives rise to the risk. An example would be provided by the company, worried about the threat of an attack on its animal experimentation laboratories by animal rights activists, which closes the laboratories and ends the experimentation. However, most things we do in life involve some degree of risk, and running a business is no exception.

Risk Transfer

If a thought is given to ceasing a particular operation, such as the animal laboratories just mentioned, that might not be a commercially sustainable decision, for the outcome is likely to be that the company has to forego the profits associated with that particular activity. A more acceptable solution might be to *transfer* the risk by subcontracting the animal laboratory operations to a third party, thereby eliminating the risk whilst retaining a lower level of profitability. The cost of subcontracting will inevitably include some consideration of the expected risk, but the overall costs might be lower, because the activity is undertaken by a specialist company.

Of course, a more common form of risk transference is to insure against a potential loss so that there is financial recompense if the risk occurs. However, full cover for every eventuality is rarely practical, whilst in some circumstances the courts and legislature restrict the ability of individuals and companies to contract out their legal responsibilities.

Risk Retention

Sometimes company managements will decide to *retain* part or all of the risk. An example of part retention is provided by a decision to insure against only some of the losses estimated to follow the occurrence of a risk. It is to be hoped that such a decision is deliberately taken after a full evaluation of the risk and potential consequences. Unfortunately, it is often the result of management neglect, caused by a failure to identify a particular risk or to assess its impact.

The decision to retain the risk, either partly or wholly, will normally be determined, on the one hand, by the organisation's ability to reduce the probability of the unwanted event happening and, on the other hand, by its ability to reduce the financial and operational impact if it does occur. This leads us naturally to the final element in a risk management strategy.

Risk Reduction

The fourth option is to *reduce* the risk through safety and security measures, procedural improvements or contingency planning. This book is largely concerned with raising people's awareness of security risks and to suggest ways of reducing these through improved security.

Fire alarms, fire doors, sprinkler systems and cut-out switches will limit the *severity* of a fire. In a similar way, good security, such as barriers, locks, safes, surveillance and alarm systems, will deter, delay and detect criminals, thereby reducing the *probability* of a criminal attack. Good controls and procedures will reinforce this protection by limiting both the *likelihood* and the *severity* of an event.

Contingency planning (see Chapter 10) may enable a company to avoid total dependency on key processes, suppliers or customers, whilst the disaster recovery planning aspect will be concerned above all to limit the extent of the impact of an unplanned event on business continuity.

MONITORING AND REVIEWING POLICIES AND PROCEDURES

The process for *managing the risks* to which most organisations are subject has successively involved identifying the risks, evaluating them and then seeking to manage or control them through the four key elements of avoidance, transfer, retention or reduction. At the end of this process, there should be measures in place that will significantly reduce the incidence of criminal activity directed against the business, whilst also curbing the impact on the business of any that still occurs.

However, it is more true than ever today that organisations are dynamic: objectives, corporate structures, products, processes, markets and personnel are constantly changing. Such changes need to be reflected in the policies and procedures that have been established to manage the security risk, leading to the clear implication that these measures must be constantly monitored and reviewed, both to judge their continued appropriateness in changed situations and to ensure that organisational changes have not undermined them or led to their neglect.

Corporate change therefore requires a constant updating of the information upon which the original security decisions were based. In addition, there should be a systematic review, or **audit**, of security measures to ensure that they remain effective agents for reducing losses, are relevant to the changed circumstances, and are not compromised by the emergence of new risks. Security audits are therefore a vital element in maintaining the effectiveness of crime prevention and loss-reduction initiatives. They complete the circle that began with risk identification.

References

Bannister, J E and Bawcutt, P A (1982) *Practical Risk Management,* Witherby & Co, London.
British Crime Survey (1998) HMSO, London.
Comer, Michael J (1985) *Corporate Fraud,* McGraw Hill, Maidenhead.
Criminal Statistics for England and Wales 1992 (1993) Command Paper 2410, HMSO, London.
Kluwer Handbook of Risk Management, Croner Publications, Kingston upon Thames, London.

3

PHYSICAL SECURITY

INTRODUCTION

The chances are that anyone who is asked to define *security* will mention two things more often than not: the uniformed guard and physical security, as defined by fences, doors, locks, barriers etc. In the UK, other popular conceptions of 'security' would also include Closed Circuit Television (CCTV) and intruder alarms, which are clearly linked closely with the more traditional forms of physical security.

The importance of physical security is to be found in how it defines boundaries to property, and then deters, detects or delays unauthorised entry. It would be very helpful if physical security measures totally *prevented* unauthorised entry, but except in an environment where high security is essential, the cost of providing this level of protection can seldom be justified. Physical security, though, does play a key role in crime prevention by placing obstacles in the way of the criminal, even if the measures in place cannot guarantee, in the final analysis, inviolability. Critically, good physical security should buy time by delaying intruders sufficiently to allow some sort of response to be mounted. Effective physical security will therefore quite probably mean that the criminal will look elsewhere for his pickings, where the difficulties confronting him are liable to be less. This is the effect known as *crime displacement*, which in this case has occurred because of *target hardening*. The crime problem has not been overcome – it has merely moved on.

From the perspective of the businessman whose physical security is good enough to achieve this displacement, the crime problem *has* been largely overcome; but even the briefest analysis will reveal that if criminals are prone to attack the least well-protected property, it is vital that one's own premises do not fall into this category.

It is the case with physical protection, as with many other aspects of security, that an effective and cost-appropriate provision will only result from the prior identification and assessment of the risks. So this chapter will spend a little time examining the risk factors which help to form the physical security measures needed to combat the threats.

One such factor is the local environment in which the premises are situated. Crime Prevention through Environmental Design (CPTED) and the concept of *defensible space*, on which so much of CPTED is based, represent new applications of environmental psychology, founded on the premise that:

> 'the proper design and effective use of the built environment can lead to a reduction in the fear of crime and the incidence of crime, and to an improvement in the quality of life.' (Jeffery, 1977)

This has been an important contribution to our understanding of physical security and will also be discussed here.

Understanding the nature of the risk and the influence of environmental factors are necessary precursors to a more detailed consideration of the range of physical security measures available. Although it has been the authors' intention to produce this book in a form that allows readers to 'dip' into it one chapter at a time, according to their concerns and interests, it is not totally possible to divorce physical security from security technology and CCTV, both of which have separate chapters devoted to them. To some extent, therefore, readers are likely to find it useful to cross-refer to the following two chapters, especially where they want to explore the technical detail of a security device or to refer in depth to CCTV. Here, we shall discuss the key features, problems and merits of fences, doors and locks, windows and security glazing. This is one area of security where there is guidance available to the non-expert in the form of a reasonably comprehensive range of British Standards. Finally, we shall also examine the contribution that lighting can make to security, briefly consider the influence of building design and explain the principle of layered security, looking at how and where it might be implemented.

IDENTIFYING AND ASSESSING THE RISK

In his excellent book on physical security, Lawrence Fennelly (1992) conclusively demonstrates how important it is to understand the *personality* of the premises you wish to protect, and this seems an ideal point from which to start the analysis of the risks to which they are likely to be subject. This personality emerges from a study of the premises over a full 24 hours, then extends to cover a complete 7-day cycle.

The average office building will be open from about 8.30 am to 6 pm, during which there is the regular traffic of those who work there, visitors, customers, delivery roundsmen, service engineers, and so on. Outside these hours, the building will be closed to the public, although one or two of the staff may arrive early or work late. During the evening and night, the building will be empty, a state liable to be ended only when the early morning cleaners arrive. Not even the cleaners are likely to put in an appearance from when the building shuts on Friday evening, right through the weekend, until it comes to life again on Monday. Public holidays will extend this period by another day or two. It can already be seen that quite different physical protection needs arise at different points in the cycle. Whereas controlling access of people becomes a priority during normal working hours, so that only those who are entitled manage to enter the premises, outside these times the security of doors and windows, the provision of adequate levels of illumination, and the elimination of any blind-spots that provide shelter from surveillance for the potential intruder assume a greater importance.

Personality differences do not stop there; they can be markedly varied according to the nature of the activities undertaken within the building and the building's actual location. A few examples will illustrate this. Two branches of the same bank might respectively be located on the High Street, immediately adjacent to the Police Station or serving an out-of-town shopping complex. One GP's practice might be established in the centre of a city and contain a supply of Class A drugs kept in a small safe; another doctor might operate a private practice in a prosperous suburb from an annexe to his home, in which he has several valuable works of art. Finally, the ground floor of one multi-tenanted office building operates as an all-night chemist, whereas another contains a quick-print copy shop that closes with the offices.

In any of these cases, viewing the property at midnight on a Saturday, when many town centres are full of young people visiting the local clubs and pubs, is almost certain to provide a very different perspective on the nature and extent of the threats to security from that obtained during working hours. Quite clearly, the risks attached to each of the premises chosen to illustrate this approach differ markedly, making it essential that the physical security is tailored appropriately. As always when endeavouring to combat the threat of crime, we are seeking to identify the ways in which scarce resources can be employed to greatest effect and on a scale appropriate to the risks. Crime prevention can be defined as

'the anticipation, recognition and appraisal of a crime risk, and the initiation of action to remove or reduce it',

and defining the personality of the building(s) is a very positive step towards achieving this.

This process of establishing the personality of a building is an important part of the wider examination of premises to determine criminal vulnerability, known as the *security survey* – a technique already covered in the previous chapter. However, it is worth ending this brief section with a reminder that a full assessment of the risks will establish, as an absolute priority, the criticality of every asset within the premises in terms of the impact its loss or compromise is likely to have on the business.

For example, one building might contain the results of several years' research undertaken by a substantial team of scientists. If this research represents the key to the future product range on which the company's prosperity will depend, this clearly will warrant a level of protection that reflects its importance. On the other hand, there may well be offices in which important, but routine tasks are undertaken, where the loss of facilities, equipment or records would likely be inconvenient, but where it should be relatively straightforward to find alternative accommodation, replace the equipment and regenerate the records. In such a case, it would be important to provide a minimum level of security, sufficient to deter vandals and the opportunist thief, but well short of the full panoply of physical security measures available.

CRIME PREVENTION THROUGH ENVIRONMENTAL DESIGN (CPTED)

The environmental approach to crime prevention and security was pioneered by Oscar Newman in his books on *defensible space* (Newman, 1972 and 1976). The concepts he developed have since been successfully practised in schools, business parks, residential areas, airline terminals, railway stations and bus depots. In this country, Dr Barry Poyner actively researched the same territory in the late 1970s and early 1980s, providing many additional insights into the positive ways in which the environment could be shaped to discourage crime and criminals (Poyner, 1983). Since then, the principles they developed have been widely endorsed and practised by police forces in many countries, and within the UK form an important constituent of police crime prevention advice. Because of the strong links with building and estate design, UK police forces commonly have a designated role of 'Architect Liaison Officer' within their Crime Prevention Departments, with a specific remit to inform architects, builders and property companies of the techniques they can use to 'design' crime out of new developments with which they are associated.

So how does CPTED work? In order to understand this, it is helpful first to define some of the key terms. *Environment* refers to the physical and social surroundings with which people interact, which for practical reasons are restricted to an area that has recognisable territorial limits. Thus, it would apply to a housing estate, an industrial estate, a shopping mall, and also to a large industrial complex. *Design* refers to the physical, social, management and law enforcement directives that are intended to encourage people to behave positively as they interact with their environment. CPTED initiatives seek to prevent certain specified crimes (**not** all crimes) and the fear they provoke, within a specified territory, by manipulating aspects of the environment that influence people's behaviour.

The CPTED approach will acknowledge the designated use of the area, define the crime problem(s) incidental to this use, and identify and implement the crime prevention strategies that will enhance its effective use. It differs quite markedly from the traditional target-hardening approach to crime prevention by drawing on behavioural theory, social science and community organisation theory, as well as good physical and urban design. In fact, it is normally possible to adapt aspects of the environment to achieve a natural form of access control and surveillance, thereby achieving a similar effect to that routinely derived from mechanical hardening and technology.

Examples of this approach can be seen now in many housing developments and trading/industrial estates. The provision of natural access control and surveillance develops a sense of territoriality among legitimate users and residents, thereby encouraging in them a sense of proprietorship, whilst at the same time discouraging potential offenders by making them aware of these subtle barriers and promoting their perception of risk. By demarcating specific spaces for specific individuals, these symbolic or psychological barriers are intended to convey to the potential criminal that unwarranted intrusion is likely to provoke protective territorial reactions from those who have a right to be there. In similar fashion, by providing good unimpeded lines of sight between buildings and into the open spaces, there is a good chance that the criminal will quickly assess that he or she is likely to be seen immediately on entering the area.

There are many other aspects of CPTED, including:

- the provision of a good level of illumination;
- the avoidance of dense, high-growing shrubs and trees that interrupt natural surveillance and provide shelter for muggers;
- good crime analysis (to understand better what types of crime represent the prime threats and to help provide appropriate countermeasures);
- control of through traffic.

CPTED principles can be applied equally well within buildings, so that, for example:

- in-store display and sales shelves are lowered to a level that improves natural surveillance, both within the store itself and from the outside;
- a good level of internal lighting is provided during store opening hours, again to improve surveillance, but also to give the impression of a welcoming environment to those outside;

staff work-stations and cash points are placed strategically to improve natural surveillance;

car parking is designed to improve natural surveillance and to increase the criminal's perception of risk.

For those whose businesses are located in areas that experience a high level of criminal activity, it would certainly be worthwhile to consult the local police crime prevention officer about CPTED. Although there are clear advantages to incorporating CPTED principles into an area as it is being designed and developed, it might nevertheless be possible to adopt some of the techniques in an established area, at reasonable cost and with a realistic chance of improving both the environment itself and the crime problem. Very many CPTED initiatives are joint ventures involving the police, town planners and local government, private estate developers and those using the areas concerned.

LAYERED SECURITY

There was an indication earlier (except in maximum security areas, like prisons, military barracks and certain classified defence-related establishments) that it was difficult to justify the costs involved in providing physical protection at such a high level that penetration became virtually impossible. The realistic objective for the average business is to provide sufficient deterrence to persuade most criminals to target somewhere else or, failing this, to provide barriers that will delay the intruder sufficiently to allow some sort of response to be organised before the crime can be completed.

At the root of this strategy lies the concept of defence in depth, sometimes called *layered security*. This is a very straightforward approach, but does require some preliminary analysis and groundwork.

Before making any decision on what kinds of physical security to install, it is vital to identify what items within a building or a complex are likely to be targeted by criminals. These will probably include cash, computers, works of art, saleable items of stock (some of which will be more valuable than others) – the choice will vary from business to business.

Some thought now needs to be given to where security devices might be ideally placed to provide the most difficult access – it does not make a great deal of sense, for example, to site a Personal Computer on a desk immediately inside a window that fronts on to a public thoroughfare, or to locate the wages office by the main entrance. Practicalities being what they are, however, it is quite likely that compromises must be made between ideal siting from a security perspective and functional convenience in general use.

Having identified the areas that are likely to be at greatest risk, the manager should now conduct a walk-through of the building, starting from the target and working towards the outside of the premises. As this is done, weaknesses in the physical barriers meant to impede unauthorised access should be noted, for it is these that should be prioritised for improved protection. Thus, a cash office handling large sums of money should be sited well within the building, and should have solid doors and frames, with adequate security locks and attack-resistant glass in any window. An intruder alarm might be required for the cash office itself, backing up one which is triggered by the initial forced entry into the building. The corridor approach might warrant both a keypad access control and surveillance from a CCTV camera, whilst there should also be good physical protection provided by the building's external doors and windows. If the building is just one within a bigger complex, then consideration should be given to the provision of a suitable perimeter security fence, but always in relation to the value of the assets that need protecting and the impact on the business of any serious breach of security. Even if the corridor approach to the cash office does not justify a camera, CCTV surveillance of the outside of the building (especially any vulnerable points) and/or the perimeter might give valuable early warning of a security incident.

In summary, layered security consists of the provision of multiple barriers to control access between areas and activities that carry a risk and the outside of the premises. It is rather like an onion skin: you peel off one layer and there is another underneath.

So far in this chapter we have been concerned with the principles and strategies that underpin good physical security, but it is now time to turn our attention to some of the component parts and look at the important aspects of doors, locks, glazing, windows, fences, lighting and safes.

As this book is not specifically aimed at prison authorities, we will continue to assume that the main threat is external (ie someone trying to break *into* the premises) and therefore we will examine the individual barriers in the order an intruder is likely to meet them. Each barrier will be examined for its most critical properties – its ability to **deter** a potential intruder or to **delay** an actual intruder. If, during the survey, an **internal** threat is identified, then it may be appropriate to use some of these barriers to delineate and protect restricted areas within the site or building.

PERIMETER BARRIERS

Walls. A high brick or concrete wall, topped with some form of anti-scaling device (eg razor-tape, spiked rollers, broken glass, anti-climb paint, etc) remains one of the most effective perimeter barriers. However, it is also one of the most expensive options and is often criticised for providing concealment for the intruder. Nevertheless, this weakness is easily overcome and it is usually cheaper to upgrade security around an existing wall than to start from scratch with some form of fencing.

Chain-link fencing. This is the most common type of perimeter fencing but has limited security value, even when installed to **British Standard 1722 (part 10)**. It is constructed of mild steel wire (galvanised or plastic coated, or both) interwoven into a diamond-shaped mesh (size 40–50mm). Without additional aids (CCTV, alarms etc), it does little to *deter* and, even with such aids, provides minimal *delay*, because cutting a single vertical strand at the top and bottom, then spiralling it upwards clear of the fence will enable that particular section to be parted sufficiently for a vehicle to pass through. Unless firmly fixed to, or embedded in, the ground, it is a relatively easy task to distort the fencing sufficiently at ground level to allow a person to crawl through.

Welded mesh fencing. This consists of horizontal and vertical strands of galvanised mild steel wire welded together at each intersection. Mesh sizes vary from 50mm square to 75×12.5mm, and with wire diameters ranging from 2.5mm to 4mm. Installed to the above British Standard, this type of fencing provides a slightly enhanced *delay* element, as it requires many more individual cuts to be made to produce a sizeable hole. Its more rigid structure also provides a more suitable support for most fence-mounted alarm systems. Complete panels can be plastic coated which, although adding to the initial purchase price, will minimise maintenance costs.

Expanded metal fencing. This type of fencing is rapidly gaining favour within the security industry. In many ways similar to welded mesh, the panels are manufactured in various sizes and thicknesses. The heavier duty panels provide quite a rigid structure and, when properly secured to the posts, will increase the time needed to penetrate by cutting.

Strands of barbed wire or razor tape used as fence topping on any of the above will make scaling more difficult and enhance the overall deterrent effect of the fence.

Steel palisade fencing. Sections consist of pre-formed steel uprights (approximately 4 inches wide and up to 10 feet high) welded or riveted to horizontal bars. The tops of the uprights are normally sharp, pointed and splayed to make scaling more difficult. The whole structure is more rigid than chain-link or welded mesh and is slightly more difficult to scale. If the horizontal and vertical strips are riveted together (as opposed to welded), prospective purchasers should ensure that the rivets are protected, since they are vulnerable to attack by small bladed chisels. If only the bottom rivets are removed, allowing the uprights to be swivelled, this breach of security may go unnoticed until the fence is physically checked. Visibility through the fence diminishes the further one strays from a 90° angle. Although more expensive to install than the wire fences, palisade fencing requires very little maintenance and is physically much stronger.

Electric and electrified fences are discussed briefly in Chapter 4 and should be considered seriously prior to making a final decision. Table 3.1 compares the merits of the main types of perimeter security barrier.

Table 3.1 *Relative qualities of perimeter fencing*

	Wall	Chain-link	Welded mesh	Palisade	Electrified	Electric
Initial cost	H	L	M	M	H	H
Ongoing cost	L	H	M	L	S	S
Deterrent value	H	L	M	M	H	H
Delay factor	H	L	M	M	L	L
Rigidity	H	L	M	H	L	L
Through visibility	L	H	M	M	H	H

H = High   M = Medium   L= Low   S = Service dependent

Gates. It goes without saying that all perimeter barriers require some means of access for legitimate users, both pedestrian and vehicular. The key requirement here is that the gate(s), in the closed position, should have the same *deter* and *delay* characteristics as the perimeter barrier. Unfortunately, all too often gates prove to be the weakest element of perimeter security, and it is therefore worth examining some of the **more common weaknesses.**

   Hinges. On hinged gates it is not uncommon to find that the gates can be lifted off their hinges, thus defeating the locking mechanism. Plates or studs welded to the hinge pins can make this more difficult.

   Locks. Gates secured by a chain and padlock offer very little resistance to a determined intruder, particularly when they are used on double-leaf gates. A sturdy locking bar secured by a close-shackle, high-security padlock is a much better option. At least one, but preferably both halves of double-leaf gates should be held firmly in place by the use of drop-bolts fitted inside the frame at the closing edge and locked in place.

   Rigidity. Gates tend to rattle in high winds and are therefore a common source of false alarms with certain types of fence sensors. Because of this, sensors fitted to gates are often desensitised, or even switched off, thus creating a 'hole' in the secure perimeter.

Powered gates. All too often the control mechanism can be reached from outside the barrier, whilst in some cases a simple key-operated switch is actually mounted externally. Rarely does such a switch present a serious obstacle to a potential intruder.

Turnstiles. Turnstiles should be full height and present no less of a delay or deterrent value than the perimeter barrier of which it forms a part. Unmanned turnstiles should consist of four wings (90° angles) rather than three wings (120° angles), because the latter permits more than one person at a time to pass through. Bottom bars or panels should be close to the ground (less than 12 inches) to prevent people from crawling under them.

DOORS

The next barrier an intruder is likely to meet is the outer fabric of the building. Provided that the walls and roof are kept in good condition, the most vulnerable points are the doors and the windows. The key aspects of these are:

Pedestrian doors. Wherever possible, glass in perimeter doors should be avoided. Solid hardwood single-leaf doors, hinged to open outwards, will usually suffice for low-to-medium risk sites. For higher risk situations, steel or steel-clad wooden doors may be necessary. *Frames* should be firmly fixed to the brickwork. Where glazing in external doors is considered essential (usually by an architect concerned about the aesthetics of the building), they should be protected during silent hours by a stout roller-shutter adequately secured.

Vehicle or plant access doors. These usually come in the form of double-leaf wooden doors, sliding folding doors or metal overhead roller-shutters. In all cases, the mechanism which secures the door to the frame or supporting structure is critical to its defensive properties. Overhead roller-shutter doors, unless secured centrally at ground level, are vulnerable to attack by levers (a crowbar applied underneath the centre of the shutter will cause it to concertina upwards, creating a gap which is large enough for a person to crawl through). Locking the drive chain or removing power from the motor does *not* disable a roller-shutter effectively.

LOCKS FOR DOORS

Mortise locks. (See Figure 3.1) These fit within the structure of the door and may be equipped with a *deadbolt* (a metal rod or bar which is moved between the locked and unlocked position by the action of turning a key or knob), or a *latchbolt* (a spring-loaded bevelled or roller bolt, which engages in the keep automatically when the door is closed), or sometimes a combination of both. As a substantial portion of the door needs to be chiselled out to accommodate the mortise lock, it should be fitted only to doors of sufficient thickness (at least 50mm is recommended). Alternatively, reinforcing plates may be fitted over the lock to provide additional physical strength.

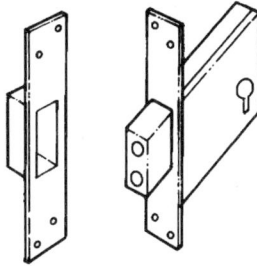

Figure 3.1 *Mortise lock*

Rimlocks. (See Figure 3.2) These are fitted to the inner surface of the door with both the lock and the keep (or striking plate) proud of the surface of the door and the door jamb. Like mortise locks, rimlocks may be equipped with deadbolts or latchbolts, or both. It is generally accepted that mortise locks (properly installed) will provide a higher degree of security than rimlocks, particularly when applied to inward-opening doors.

Both mortise locks and rimlocks used for security purposes should comply, as a minimum, with the *British Standard for thief resisting locks, BS 3621.* For inward-opening doors, it may be necessary to fit two such locks on the opening edge in order to provide adequate resistance to an attack using brute force (shoulders, feet, sledgehammers, etc).

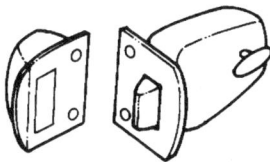

Figure 3.2 *Rimlock*

Mechanical combination locks. (See Figure 3.3) These are usually manufactured as rimlocks but, as the name implies, the bolt action is released by depressing a number of 'buttons' in the correct sequence. Although some such locks can provide good physical strength, their main weakness lies in the fact that, over time, the combination becomes known to unauthorised personnel, either due to poor discipline or to *shoulder surfing* (non-code-holders watching and memorising the numbers sequence). Although offering relatively low security, combination locks do have the advantage of enabling frequent changes of combination, which is much cheaper than changing a lock or a cylinder.

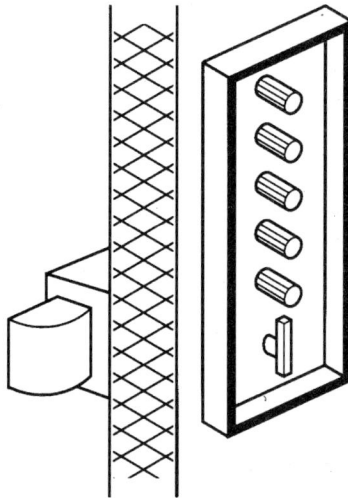

Figure 3.3 *Mechanical combination lock*

Electric locks. These come in two types: the first where the bolt is operated by a solenoid or motor; the second where the latch is released by applying or removing electrical power. The first is normally the more secure, but both suffer from the need to protect the power source and the associated cabling.

Multipoint locks. (See Figure 3.4) These locking devices are ideal for fire-exit doors which form part of a secure perimeter. They come in various shapes and sizes to fit most single- and double-leaf doors, and some are designed to resecure automatically after closure.

Padlocks. These should not be used in place of mortise or rimlocks. Good quality padlocks (ie those made from hardened steel, close shackled and with at least five levers) do, however, have a valid role to play.

Used *internally* they can be most effective in securing sliding doors or roller-shutters. Used *externally* the hasp and staple elements should provide good resistance to attack by levers, bolt croppers, or hammer and chisel.

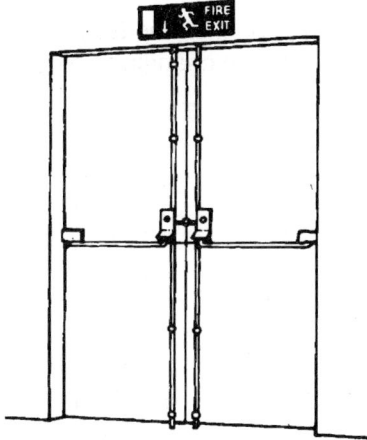

Figure 3.4 *Multipoint lock*

Monitored locks. Many security locks are fitted with internal switches which, when wired to the alarm system, will indicate whether the bolt is in the locked or unlocked position. Used in conjunction with a door contact, a *'locked-shut'* status can be monitored. Alternatively, the switch can be mounted in the keep and operated by the bolt moving to the locked position. Unfortunately, in the latter case the switch is often easily jammed or wedged in the closed position (either accidentally or deliberately).

Hinge bolts. (See Figure 3.5) Sometimes called **dog-bolts**, these are fitted to the hinged edge of the door. As the door closes, the bolts locate into holes in the frame. Hinge bolts are essential for outward-opening hinged doors because the removal of the exposed hinges will defeat the locking mechanisms.

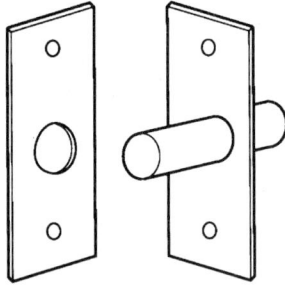

Figure 3.5 *Hinge bolt*

KEYS FOR DOOR-LOCKS

Yale keys. Named after their inventor, these are probably the most commonly used type of key (see Figure 3.6). The vertical undulations on the shank of the key raise the spring-loaded pins within the cylinder to a predetermined level, thus allowing the cylinder to turn and operate the bolt. The number of possible differs (combinations) is increased by varying the shape and size of the horizontal grooves along the shank, which allows only the correct shape of key to enter the cylinder. The facility to change the cylinder without the expense of replacing the whole lock is a major advantage. Unfortunately, cylinder locks are not the most difficult of locks to pick and are susceptible to the entire cylinder being ripped out of the lock.

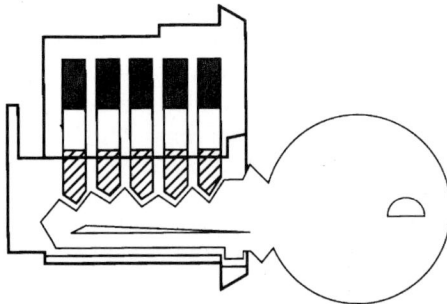

Figure 3.6 *Yale key*

Lever locks. (See Figure 3.7) These can provide a higher degree of security than cylinder locks. The key is shaped to move a number of levers into a position which enables the bolt to move. The higher the number of levers, the greater the number of possible differs. By adding **wards** (obstructions within the body of the lock which prevent the wrong-shaped key turning), the number of differs can be increased even further. If **key-suiting** (configuring the locks to enable the use of master and sub-master keys) is to be utilised, then a set of locks with a large number of differs will be required, because suiting drastically reduces the available combinations.

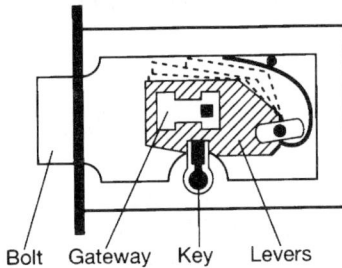

Bolt   Gateway   Key   Levers

Figure 3.7 *Lever locks*

WINDOWS

Windows are a favourite entry point for burglars for a number of reasons:

- They are often left open.
- In many instances, one or more complete panes of glass can be removed from the outside, thereby overcoming the risk of noise or injury caused by climbing through broken glass.
- Window catches fitted by the manufacturer are generally easily defeated.
- The items targeted by the villain are often visible through the window which, in turn, offers the most direct route to and from those items.
- There are often windows at ground-floor level in secluded parts of the building perimeter.

Because of these weaknesses, there is much to gain from the occupier making entry through windows just as difficult as through doors. The two most important characteristics of a *'burglar-resistant'* window are the strength of the materials from which the window is constructed and the locking device used to secure it. The more common types of windows are nowadays popularly referred to as:

Casement windows. These are side-hinged windows and may be constructed from metal, wood or plastic.

Transom windows. These are similar to casement windows but hinged along the top or bottom edge.

Sash windows. These are usually vertically, but sometimes horizontally sliding windows, which are generally found in older buildings and made of wood.

Centre-hung windows. These pivot around a central horizontal axis and are normally secured with a spring-loaded latch bolt. High-level windows of this type are reached by the use of a long pole with a hook attached to one end.

The required physical strength of the window will depend, to a very large extent, upon the risk evaluation established at the survey stage (see Chapter 2), but there are some general rules which considerably enhance the security value of most windows, as follows:

1. Try to ensure that **fixed, non-opening windows** are installed (especially at ground-floor level or where the window is accessible from the roof of an extension or garage). If opening windows are deemed essential for ventilation purposes, then use high-level transom windows which are too small to admit even a very tiny person.

2. If vandalism is identified as a major problem, consider installing a **protective outer shutter** which can be removed during normal working hours. Decorative fixed grilles can serve as very effective barriers whilst enhancing the aesthetics of a building, but they must *not* be fitted to windows which could conceivably be required as alternative emergency exits.

3. Replacing normal float glass with – in ascending order of resistance to physical attack – **toughened glass** (as found in cars), wired glass, double glazing or laminated glass will increase the delay factor appropriately. (Laminated glazing is discussed in more detail later in this chapter.)

4. Adding **mirrored transparent plastic film** to existing glass may help to obscure the potential intruder's view, provided that light levels inside are lower than those outside. Such film may also fractionally improve resistance to physical attack and give added protection against bomb blast, depending upon the film's thickness.

LOCKS FOR WINDOWS

Locks for windows are surprisingly inexpensive and relatively easy to fit retrospectively. Various crime surveys have indicated that a significant proportion of illegal entries are effected through windows at the rear of premises, thereby demonstrating that fitting window locks is certainly a cost-effective method of improving security (see Figure 3.8).

Most window locks secure the moving section to the frame with a screw or locking bolt. Some are designed to secure the window catch, but in most cases these should be avoided, as the lock will only be as strong as the catch.

Keys for the locking mechanisms vary from a simple Allen key to warded keys which give extra security. Windows are often 'jemmied', so long-throw bolts and stout fixing screws are advised. For more vulnerable windows, two such locks can make life quite difficult for the burglar.

It is important to remember that the keys to the lock(s) should be readily available (although not within reach of, or viewable through the window) if the window is to be considered an alternative emergency fire exit.

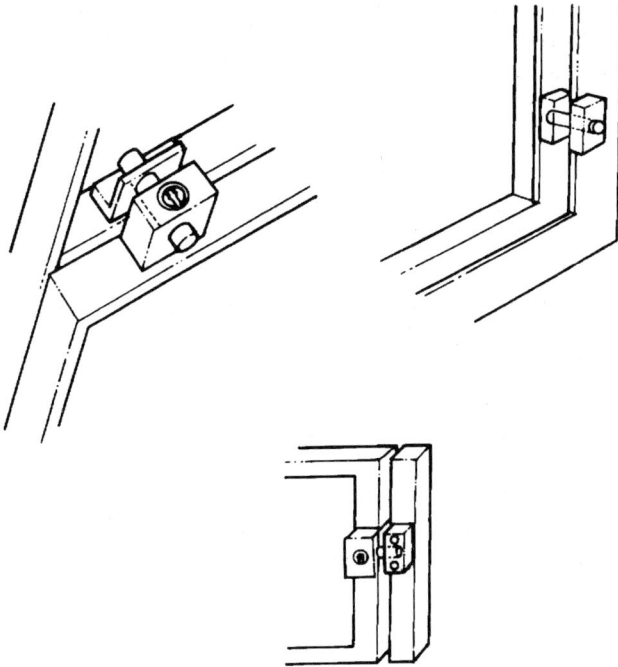

Figure 3.8 *Some of the locks available for different types of windows*

ROOFS

The roofs of buildings can be particularly vulnerable as entry points for a variety of reasons:

- They are often ignored during the survey stage.
- Slates or tiles are easily removed without creating undue noise.
- Rooflights are rarely secured or grilled.
- Access to the roof can often be gained from adjacent buildings, via materials stacked against the wall, or because ladders have been conveniently left lying around.
- Buildings are rarely alarmed to detect attempted entry through the roof.

Unless they are required for ventilation purposes, **rooflights** should be non-openable and constructed of glass-reinforced polyester (GRP). Where ventilation is deemed essential, such apertures should be fitted with bars or grilles to delay entry attempts (see Chapter 4 for the appropriate alarm devices).

ACCESS WITHIN THE BUILDING

Control of access to specific areas within a building is particularly important where the general public are admitted. Generally speaking, it is not necessary to install walls, partitions, doors or windows to the same high-security specifications that apply at the perimeter of the building. Mechanical combination locks, or electrical locks operated by personal codes or access control cards and fitted to standard doors will normally suffice in maintaining a *'need-to-be-there'* principle.

One exception to this arises where the general public interfaces with employees who have direct access to substantial amounts of cash or valuables (eg a bank or post office counter). In such cases there is an additional risk, which is that an employee or customer might be injured or worse. There are three main schools of thought as to how best to tackle this problem:

1. Erect a **physical but transparent barrier**, which allows customers and staff to see and converse with each other, but which will protect the employee from threats of physical violence, even from the use of firearms. The main advantage of this approach is that it offers a permanent barrier behind which the employee is protected whilst taking evasive action, or operating a raid alarm button. Cost is a significant factor with this option, since the entire barrier, both the transparent and opaque sections, need to be constructed to the same high specification, as do any doors, drawers or transfer hatches

which form part of the barrier. Despite the latest electronic technology, or acoustically enhanced frames or hatches, voice attenuation remains one of the primary drawbacks to this type of installation.

2. Maintain an **open-plan office approach** where there are no physical barriers between customers and staff, whilst ensuring that the employee does not have access to more than a token amount of cash or valuables – *and clearly advertise this fact*. This approach is favoured by the sales and marketing fraternity who perceive it to be *'user friendly'* and more attractive to customers. It does little, however, to protect staff from frustrated robbers who are increasingly becoming more violent.

3. Erect a waist-height barrier (counter) which contains a **rising screen**. Such a screen would normally consist of a bullet-resistant, opaque panel which, on triggering, will rise to the ceiling within a few hundred milliseconds, thereby cutting off all forms of communication between the customer (villain) and members of staff. It is argued that the sudden insertion of this barrier to spoken and visual communication, and to physical access, isolates the attacker from any intended target so completely that there is no alternative but to abort the attempted robbery. There would also be no point in taking an innocent bystander hostage, makers claim, as there is no one left to threaten.

SECURITY GLAZING

This is an appropriate point at which to examine the relative merits of security glazing. Earlier, we mentioned some of the glazing methods used for perimeter windows, but if we wish to seriously delay penetration by a determined intruder or to protect staff from armed robbers, then we need to consider much stronger materials. Such products tend to fall into two main categories: *anti-bandit glazing* and *bullet-resistant glazing*.

Anti-bandit glazing – British Standard 5544. This type of glass defines the expected performance levels against non-ballistic weapons (eg attacks using sledgehammers, pickaxes, etc). Anti-bandit glazing is normally manufactured from one of the following:

Laminated glass is made from sheets of float glass that are laminated together using polyvinyl butyral (PVB) as the adhesive interlayer. Resistance to physical attack rises as a function of the thickness of the glass and the number of layers. An overall thickness of 6–7mm is recommended for shop-front windows, and 11–12mm for protective screens designed to prevent physical penetration.

Plastics. A number of thermoplastic materials, most notably acrylic and polycarbonate, can be moulded into almost any shape to produce structures which can withstand severe impacts. On an equivalent strength basis, plastic products are much lighter than glass/PVB laminates and generally less expensive to install. They do suffer, however, from damage due to scratching and other abrasions which shorten their useful lifespan, particularly where transparency is an important feature of the structure. Modern products boast '*scratch-resistant*' surfaces, but are still susceptible to defacing. Most plastic screens are vulnerable to heat and therefore do not offer much resistance to a naked flame.

Composites. Glass and plastics can be laminated together to form a screen which combines the impact-absorbing qualities of plastics with the inherent scratch-resistant property of glass. Some manufacturers claim to have overcome the problems initially encountered with this technique – namely, the manner in which the glass and the plastic interlayers expanded and contracted at different rates with changes in temperature, and the inclination of the plastic and adhesive to deteriorate if exposed to direct sunlight. With some designs the glass and plastic layers are separated by an air gap, with each layer being held firmly in place by the frame.

Bullet-resistant (BR) glazing – British Standard 5051. This type of glass defines the expected performance levels against attackers using firearms. The standards are very precise, specifying the defensive properties required to withstand an attack from firearms that range from a 9mm handgun up to a 7.62mm rifle and a 12-bore shotgun, all fired from a set distance, at a minimum velocity and at a 90° angle. Such parameters are rarely found in live situations, but it is always safer to plan for the worst scenario. Most BR glazing panels are of a glass-laminate construction similar to that of anti-bandit glazing, *but much thicker and therefore much heavier*. The attack side of the panels is designed to dissipate as much energy as possible, whereas the protected side is designed to minimise spall (pieces of glass flying off). It is important, therefore, to install the panels the right way round. **Composite** panels can provide an equivalent degree of protection, are generally much lighter and often compare favourably on price.

We have now considered the major vulnerabilities of the building perimeter and areas within the building. Items of significant value will require a degree of physical protection much greater than that afforded by fences or walls.

SAFES AND STRONGROOMS

The next and, without doubt, the most effective, layer of physical security in our *'onion skin'* strategy is the **safe, strongroom or vault**. Such an enclosure, constructed to the proper specifications or standards, will present the would-be thief with a very difficult obstacle. Although its presence automatically *deters* all but the most determined or optimistic villain, its main virtue lies in its ability to *delay* access to its contents.

Safes come in various shapes and sizes, from small wall or under-floor models through to large double-door safes of almost 'walk-in' proportions. The size and the weight are important features because impenetrability would be of little value if the thief could simply remove the entire enclosure to a remote location where time, or the availability of appropriate tools, would no longer be a limiting factor in opening the safe. In most instances (and certainly if the safe weighs less than half a ton), the safe should be securely bolted to the floor or wall (or both), encased in concrete, or otherwise immobilised.

The next factor to consider is the **physical construction** of the safe. A safe's quality (its resistance to being penetrated) is measured by the time it takes to gain access to its contents using a particular set of tools or devices. Many countries have their own standards of measurement (some have none!), but probably the best known are those of the Underwriters Laboratories Inc in the USA. Without delving too deeply into the technical specifications of UL standards, those for safes broadly fall into three categories:

1. **TRTL 30:** will resist attack by hand-held mechanical and electrical tools, oxyacetylene torches, etc for a minimum of 30 minutes.
2. **TRTL 60:** as for TRTL 30, but will resist penetration for 60 minutes.
3. **TXTL 60:** as for TRTL 60, but adds high explosives to the list of attack materials.

Such safes use an amalgam of heat- and drill-resisting materials, designed to resist attack on the body or on the door of the safe.

The **safe door** is the most complex part of the safe. Its design and construction must allow legitimate access to authorised personnel whilst denying access to others. When locked, its resistance to physical attack must be commensurate with the other five sides. To achieve this, a high-quality safe will have the following characteristics:

   Locking bolts will penetrate the main body of the safe at the top, bottom and closing edges of the door.
   Hinges will be protected and hinge bolts or a fixed blade on the hinged edge of the door will engage into a recess in the body of the safe as the door closes.

Relocking bolts will activate if explosives are used in an attempt to open the safe.

Multiple locking devices will be fitted (see p 43).

It is worth mentioning that many safes in use today were purchased several years ago and are no longer capable of withstanding an attack in which modern tools are used. If in doubt about the quality of an existing safe, you should always consult your insurer.

Strongrooms are primarily 'walk-in' safes and should exhibit similar security characteristics. As with safes, the UL standards are based upon the time taken to penetrate the structure (Class 1 = ½ hour; Class 2 = 1 hour; Class 3 = 2 hours). Strongrooms tend to be constructed in one of three ways:

1. **In-situ**. These are constructed from the ground up by pouring concrete over a variety of reinforcing structures. This is a specialist's job, as the quality of the concrete mix and the positioning of the reinforcing materials are critical to the walls' defensive properties.
2. **Pre-cast blocks.** Concrete blocks (or beams) are factory manufactured and transported to the customer's premises, where they are assembled on site. The blocks or beams are interlocked (usually with steel dowelling rods) to form a continuous enclosure, leaving a suitable aperture for the door.
3. **Demountable.** Sometimes referred to as **portable** strongrooms, these consist of pre-formed panels which are manufactured from a variety of materials, including high-density concrete, heat- and drill-resistant materials, and then bolted or welded together on site. Demountable strongrooms have two significant advantages over their more traditional counterparts:

   — They are lighter and therefore more amenable to installation in the upper floors of high-rise buildings.
   — They can be reduced or extended in size as requirements change.

   But beware: the name implies that they can be deconstructed and reassembled easily and at minimum cost. However, the author has experienced instances where such an operation has been almost as expensive as buying a brand new strongroom.

Subterranean strongrooms are often referred to as **vaults**. They should be constructed to the same standards as their surface cousins, but they have a major advantage in that four of the six sides are surrounded by earth, which denies the villain space in which to operate when attacking the structure. One should have the surrounding substratum surveyed carefully, however, as sewerage systems and the like may provide a concealed approach path to the vault.

LOCKS FOR SAFES AND STRONGROOMS

There are three basic locking systems for safe and strongroom doors. All three fulfil the same function – in the locked position they inhibit the movement of the locking bars within the door. Taking each in turn:

1. **Key locks** are not dissimilar to high-security locks for ordinary doors. They are embedded deep into the door, behind the main protective layer, and are operated by keys with long shanks. Key locks have two drawbacks:
   — Most keys are fairly easily copied.
   — There is always the problem of protecting the key after the door has been locked.
2. **Combination locks.** Modern high-security combination locks are reliable and effective. Combinations can be committed to memory and are easily changed if compromised. The main weakness with combination locks lies with the combination holder. It is not unusual to find the code *'hidden'* in a desk drawer or written down on the back of a wall calender. Code holders are also prone to sharing the codes with *'trusted'* colleagues (just in case they are late for work). *'Shoulder surfing'* is another weakness of combination locks. Despite all this, if combinations are changed regularly (and immediately if a code holder leaves), they remain the preferred locking mechanism for strongroom doors.
3. **Timelocks** are designed to prevent the safe or strongroom door being opened within a predetermined time window, even when all other securing devices are unlocked. A timelock is a major safeguard against forced entry during non-working hours. Timelocks are manufactured either as *mechanical* devices (normally two or three clockwork movements, which are wound up to run for a selected number of hours), or *electronic timing* devices, which can be pre-programmed to enable or disable the locking mechanism at set times. Although the electronic timelocks offer a much greater degree of flexibility, they have yet to prove that they are as reliable as their mechanical counterparts.

   It is strongly recommended that an external indicator, which monitors the state of the timelock, is mounted adjacent to the door (mistakes are sometimes made in setting the timer and it is nice to know, when the door won't open, that the timelock is the cause!).

   Time-delay locks are designed to thwart robberies during working hours. These locks prevent the door being opened until *'n'* minutes (usually one to ten) after the device is triggered – the theory being that the robber will not wish to hang around that long. Unfortunately, most of these devices are mounted externally and some are not very robust and may succumb to a heavy blow from a sledgehammer.

We have now examined most of the physical barriers which a villain should have to overcome in order to reach the protected assets. It would be inappropriate to leave this section on physical security without mentioning **security lighting**.

SECURITY LIGHTING

Illuminating a site or building perimeter will not *delay* an intruder, but an intelligently designed installation may increase the *deterrent factor*. If it is used in conjunction with CCTV (see Chapter 5), good lighting will certainly assist in *detecting* unwelcome visitors. There are many factors to consider when planning a strategy for security lighting. Here are some of the most important:

Which areas need to be illuminated? The key areas for consideration are:

- the site boundary or perimeter (including external car parks);
- the open areas between the perimeter and the building(s);
- pedestrian and vehicle access points;
- the building interior.

Which type of lighting should you use? The more common types can be divided into three groups – filament lamps, fluorescent lamps and high-intensity discharge lamps. Their main characteristics are:

- **Filament lamps** are of two main types in common use:
  - *Standard tungsten*, which are commonly used in most households and are low-cost, low-efficiency light sources with an expected lifespan of around 1000 hours. They are normally limited to indoor use or to supplement the main external security lights.
  - *Tungsten halogen* are more efficient than standard tungsten and last around 2000 hours. They have a wide frequency spectrum, which makes them ideal for use with CCTV systems.
- **Fluorescent lamps** are more efficient than tungsten bulbs and have a considerably longer lifespan (5000 hours +). These lamps are limited by their low-output power and are normally used indoors.
- **High-intensity discharge lamps**. Those most commonly used for security lighting are described below:

— *Low-pressure sodium (SOXE)* are very efficient in terms of light produced per watt consumed, and have a relatively long lifespan (approximately 8000 hours), but emit a yellow light which, due to its very narrow frequency spectrum, is unsuitable for areas which are under CCTV surveillance.
— *High pressure sodium (SON)* have a slightly longer lifespan but are less efficient than SOXE. They boast a wider frequency spectrum and, although still too narrow to produce good quality images from some **colour** CCTV cameras, the DeLuxe version with an extended white light range does improve colour rendering appreciably.
— *High pressure mercury vapour* are less efficient than SON, but have a similar lifespan. These have now largely been superseded for security purposes.
— *Metal halide lamps* are similar to mercury vapour, but more efficient, and have excellent colour rendering which makes them ideal for external CCTV applications.

Continuous or interrupted use? It is important to remember that discharge lamps can take several minutes to strike and get up to full operating brightness, whereas filament and fluorescent lamps strike the moment power is applied. On the other hand, the latter also fail instantly, whereas the former tend to lose power over a protracted period. It follows that discharge lamps, having higher efficiencies and longer lifespans, are more suited to perimeter and area lighting. The drawback of slow restrike rates can be partially alleviated by connecting alternate lamps to separate phases of the mains supply, although this doesn't help much in the event of a total power failure. The ideal system will use a mixture of lamps.

Positioning of lamps? Ideally, the lighting installation should bathe approaching intruders in light, whilst leaving the defenders in the shadows. Such a system is not always appreciated, however, especially by neighbours or passing motorists, so some form of compromise is usually necessary. Light-fittings can be equipped with various types of **lenses** to shape the light into the required pattern. The lamps can be mounted on posts along the perimeter, on high masts or on the sides of buildings. This last option has the advantage of providing a degree of protection to the power cables. Perimeter lamps may need to be fitted with toughened glass lenses or wire mesh guards if vandalism is deemed to be a problem.

Automatic operation? The two most common methods are:

A light-sensitive device that switches the lamps on automatically as daylight fades, thereby avoiding the risk of lights being left switched off.

External movement detectors (see Chapter 4) that can be used very effectively in conjunction with spot- or flood-lights, but which *must* be of the 'instant strike' variety.

SUMMARY

In this chapter we have examined some of the physical barriers which can be used to *deter and delay* a potential intruder. Of all the constituent parts of a sound loss-prevention strategy, good physical security is without doubt the most important. However, a strongroom constructed to the highest standards cannot be expected to protect its contents if the custodian goes home without locking the door – so *management procedures* to ensure the integrity of the physical security are equally important.

References

Fennelly, L J   (ed) (1992) *Effective Physical Security*, Butterworth Publishers, Stoneham, Massachusetts.

Jeffery, C R (1977)   *Crime Prevention Through Environmental Design*, (2nd edn), Sage Publications, Beverly Hills, California.

Newman, O (1972) *Defensible Spaces: Crime Prevention Through Urban Design*, Macmillan, New York.

Newman, O (1976) *Design Guidelines for Creating Defensible Space*, Law Enforcement Assistance Administration, Washington DC.

Poyner, B (1983)   *Design Against Crime: Beyond Defensible Space*, Butterworth Publishers, London.

4

SECURITY TECHNOLOGY

INTRODUCTION

With increasing frequency directors or managers of companies find themselves conducting *post-mortems* into the cause and effects of burglaries, robberies or wanton acts of vandalism. Forcible entry to premises has become a boom industry and visits to some scenes of crime by police are no more than perfunctory, which is not surprising as police figures show that the false alarm rate from automatic intruder detection systems is still running at well over 90 per cent. Unfortunately, all too often a high-pressure salesperson from an alarm company will try to sell security equipment to a manager or director who may know what the problem is, but isn't quite sure what is required to solve it. Frequently the result is a very expensive, poorly designed security system, which is more effective at producing middle-of-the-night false alarms than detecting intruders.

Large companies can afford their own in-house experts or external consultants, but smaller companies tend to rely upon the general advice of the local Crime Prevention Officer and the more detailed technical advice of the alarm company representative.

As most of the relevant technology involves electronics and computers, it would be very easy to over-complicate some of the issues involved in designing and installing effective security systems. Here we will attempt to examine the more common elements of a good defensive system, at the same time pin-pointing some of the strengths and weaknesses of the various devices.

In Chapter 3 we discussed how various layers of physical security could enhance the **deter and delay** elements of our **'three ds'** strategy. In this chapter we will examine how intruder alarm systems can provide the **'detect'** element in a reliable and cost-effective manner. We will also consider some of the latest technical developments in alarm verification techniques (to reduce false alarms), automatic access control systems and other relevant technologies.

INTRUDER DETECTION SYSTEMS

It is important to remember that intruder detection systems (usually referred to as alarm systems) do not, in themselves, prevent your assets from being stolen. They merely inform you that someone is stealing, or attempting to steal, that which you are trying to protect. In the main, businesses use alarm systems to help protect their premises during non-working hours, having identified external criminal elements as the main threat.

One interesting definition of the 'perfect' alarm system is 'a system which can reliably detect the presence of a human being with criminal intent'. Now, we have not yet come across a system capable of distinguishing between people with, or without, criminal intent, so for now we will just have to settle for systems which detect the presence of humans. Of course, if we have designed our physical security adequately, one could argue that only those with criminal intent would be present within the detection area, thus satisfying the definition.

In the previous chapter we saw how physical barriers should provide the time-delay element so crucial to our strategy. Equally important is the ability of an alarm system to detect intruders or would-be intruders at the earliest opportunity and to signal this fact to some form of reactionary force, be it police, private security company or, indeed, members of the management team. Only if this reaction time is shorter than the delay caused by our physical barriers will our defensive strategy be successful.

A good alarm system will satisfy four key requirements:

1. To detect reliably the presence of humans (with criminal intent?).
2. To detect reliably penetration of physical barriers.
3. To provoke an immediate response to an activation.
4. To be free of false (or nuisance) alarms.

Of course, it is also important that the cost of the system is appropriate to the value of the assets being protected. The installation and maintenance costs are particularly relevant when comparing systems of similar specifications.

An alarm system can usefully be considered as consisting of four separate but interdependent functions:

1. *'Risk-end' detection.*
2. *Control and monitoring.*
3. *Signalling.*
4. *Response procedures.*

Remembering that our prime objective is to detect, as early as possible, a threat to our assets, then to provoke a speedy and effective response, we shall look at each of the four elements separately.

RISK-END DETECTION EQUIPMENT

These are the devices which should detect the presence of humans or the penetration of a physical barrier and signal the event to the control and monitoring equipment. In designing a system, it is often best to start at the perimeter and work inwards, remembering that the enemy is usually external and the object of the exercise is to raise the earliest possible warning that our physical security is under threat.

Perimeter Alarms – Outdoors

The more common technologies will be reviewed in two groups: **free-standing systems** and **fence-mounted systems**.

Free-Standing Systems

Infra-red beams

These are best imagined as a taut piece of invisible string stretched between two points. A series of horizonal beams are transmitted between towers which covertly house the transmitters/receivers. By analysing the sequence, number and duration of beam interruptions, it is possible to provide some degree of discrimination between real and false alarms. Infra-red (IR) beam arrays are expensive but reliable, although the longer-range arrays tend to suffer from attenuation in heavy rain and fog. IR beams would normally be positioned just inside the perimeter fence but far enough from it to prevent intruders using the fence posts as platforms to jump clear of the beams.

Microwave 'fence'

Overlapping transmitters and receivers provide line-of-sight beams of microwave radiation. Objects moving within these beams will alter the pattern of radiation arriving at the receiver. Arguably more effective than IR beams, it is vital that the ground between the transmitter and the receiver is kept free of obstructions, including long grass. The main disadvantage of microwave 'fences' is that they require a substantial area of clear ground around the perimeter and are prone to false alarms from animals, debris, etc. As with IR beams, they should be positioned inside the boundary fence or wall.

Ground-pressure systems

These systems consist of buried fibre-optic cables or fluid-filled conduits which detect the pressure differentials caused by people or vehicles moving over the protected area. Although expensive to install and maintain, they are reasonably effective and fairly free from false alarms.

Fence-Mounted Systems

Microphonic (or acoustic) cable

This technology is now very popular due to its relatively low cost. Best imagined as a long microphone (up to 300 metres or more), the cable detects the vibrations and noise created by cutting, climbing or tunnelling under the fence. The signals are then processed electronically in an attempt to differentiate between 'normal' sounds and those produced by an intruder.

Vibration sensors

These usually consist of inertia or piezoelectric devices connected to the fence or posts, or both. Inertia-based sensors form part of an electric or electronic circuit which is interrupted when the fence is disturbed. Piezoelectronic devices perform the same function by generating electrical signals when disturbed. These signals are detected, analysed and compared to the pattern of signals expected to be generated by an intruder.

Fibre-optic cables

These can be utilised in two ways. The simplest and most reliable (if not the most effective) system is where the fibre optic is an inherent part of the fabric of the fence. Detection is therefore dependent upon an intruder severing the fibre optic. A more sophisticated but more costly option is where the light travelling along the length of the cable changes pattern slightly in sympathy with the physical movements of the cable – this slight change is detected and evaluated by optical sensors. In many ways similar to the microphonic cable, this configuration is less prone to false alarms (particularly those due to electromagnetic interference), although it is more costly and arguably less sensitive.

E-field (or capacitance)

These are conductive cables mounted on, but insulated from, a fence or wall. They form part of a resonant circuit, the frequency of which is disturbed by the capacitance effect of a body entering the electromagnetic field. Although effective at detecting intruders, the adjacent area needs to be kept free of foliage and debris to prevent false alarms.

Other Fences

Two types of fence which embrace both physical and electronic protection are worth noting:

Electrified fence

This is a development of the system used for controlling farm animals. Whether installed as a complete physical barrier or mounted on an existing barrier, the theory of operation is the same. Momentary pulses of *very high voltage* (but low current and therefore safe energy levels) are transmitted along the wires, providing a nasty shock to anyone attempting to scale the fence. The system will also detect climbing and cutting attempts, but its main strength lies in its deterrent value. To enhance this feature, prominently appointed warning notices are attached to the fence at regular intervals. This system is particularly suitable when the property boundary represents the sole physical barrier, eg car showroom forecourts, scrapyards, etc. If mounted on existing structures, care should be taken to ensure that the electrified fence does not protrude beyond the property boundary.

Electric fence

This consists of tensioned strands of barbed wire, insulated from the supporting posts and conducting *low voltage electricity*. Cutting or shorting the specially arranged strands will cause an alarm. Provided the fence line is kept clear of foliage and debris, this type of fence system boasts a particularly low rate of false alarms.

Avoiding Problems

Passive infra-red detectors, known as PIRs (which are discussed later in this chapter), and video motion detectors, commonly abbreviated to VMD (and discussed further in Chapter 5), are also used occasionally as external devices. However, their use on unmanned sites as primary response-generating systems is not recommended, due to the high incidence of false alarms.

Many other, less frequently used, systems are available, but all seek to achieve the same aim – to detect penetration into the protected area. Most seek to achieve this in one or both of two ways: to detect the physical presence of humans and/or to detect damage to the physical barrier. Remembering that the twin objectives are to maximise the detection capability and to minimise false alarms (very prevalent in external alarm systems), it is worth emphasising some common problems. These are:

- If the perimeter fence is adjacent to a frequently used public area, it is usually prudent to avoid fence-mounted systems. In this instance a free-standing system within the perimeter is preferable.
- Extremes of weather, vegetation growth and animals are all major causes of false alarms.
- If used as a first line of defence in a 'lock-up-and-leave' situation, it would be wise to consider:

  — installing one of the more stable systems (unfortunately this usually means losing some detection value);
  — if the risk is great enough to warrant it, combining two systems (which detect two different physical phenomenon) and configuring the alarm outputs to 'handshake', ie both systems must activate within a given period for an alarm signal to generate a response;
  — using one of the alarm verification techniques (discussed later) in conjunction with the perimeter alarm.

Perimeter Alarms – Indoors

Often a business will not have the luxury of sufficient space for a perimeter fence or wall so that the outer fabric of the building forms a first line of defence. However, the strategy should remain the same as before – to detect intrusion or, preferably, attempted intrusion. However, in these circumstances, the methods of execution will vary from those we have just considered. Accordingly, in the sections that follow we shall examine devices that are designed to detect entry through doors, windows, walls, floors and roofs.

Doors

The most common form of device which detects the opening of doors is the magnetic reed switch. These can be installed either as flush mounted (the switch recessed into the frame, the magnet into the door) or surface mounted (where both the switch and the magnet are visible from the protected side). The former is slightly more expensive to install but is aesthetically more pleasing and fractionally more secure. These switches are reliable and inexpensive but do not present serious difficulties to a knowledgeable intruder. As discussed in Chapter 3, external doors should be fitted with a good quality mortise lock. It is a good idea to fit a monitored keep (a switch within the keep that is operated when the bolt enters the keep) because when this is wired in series with the door contact, it ensures that the door is both closed and locked (otherwise the alarm system will not set).

Penetration through the fabric of the door can be detected using closed circuit wiring (either copper wires or fibre optics covering the inner surface of the door), but this is expensive and liable to damage if it is not adequately protected. Volumetric devices protecting the area inside the door are usually a more viable option and these are discussed below.

Another, rather out-of-fashion but nevertheless effective, device is the pressure pad. Normally placed directly behind the door or beneath windows, and concealed below the carpet or vinyl floor-covering, these pads are designed to produce changes in resistance (or short/open circuits) in an electrical circuit when someone stands on them.

Windows

Strips of foil which, when broken, interrupt an electrical circuit, used to be a common method of protecting large window panes. However, they are prone to damage (usually by window-cleaners) and although providing a visible deterrent, are easily overcome. This device is also unsuitable for windows with multiple small panes.

'Break-glass' detectors are now more common than foil and come in two basic forms:

- *Piezoelectric devices* which are attached to the surface of the glass. The vibrations created by breaking the glass generate electrical currents which are detected and analysed. For best results, each pane requires its own detector.
- *Acoustic detectors*. These are designed to detect the narrow band of frequencies which are generated when glass is broken. An analyser differentiates between these frequencies and those normally present in the surrounding environment.

Properly installed, both types of detector are reasonably effective and should not be prone to excessive false alarms.

Of course, break-glass detectors do not detect windows being opened, so openable windows need further protection, such as magnetic reed contact switches. For higher risk areas, internal grills should be considered; these can be fitted with vibration sensors set to quite a *low* sensitivity level, otherwise unwanted false alarms might well result.

Volumetric Space Protection

Although it is eminently sensible to detect penetration through doors and windows – these are, after all, the points of entry most often used by intruders – one must remember that buildings and other structures usually have six sides, with roofs particularly vulnerable to attack. It would be very expensive to provide perimeter protection for the walls, roof and floor, so the obvious answer is volumetric (or space) protection. These devices are designed to detect a person or persons entering or moving about within the protected area. The more common detectors within this category are considered next.

Passive infra-red detectors (PIRs)

These are probably the most common form of detector in use today. Advances in IR and optical technology have resulted in PIRs becoming effective and reliable movement detectors. They work by detecting infra-red radiation generated by normal body heat. An array of optical lenses within the device focuses the radiation on to a sensor, the output of which is modulated by the source of heat moving from one 'optical' zone to another. Intelligent arrangements of the arrays and subsequent processing of the signals minimise false alarms.

The lens configuration is also critical in determining the size and shape of the volumetric area to be covered. PIRs tend to come in three types:

- **Volumetric.** Wide-angle, three-dimensional cover for open areas such as offices, factories and storerooms, with a typical range of 10–20 metres.
- **Long range.** Narrow angle, long-range beams for corridors, gangways and between storage racks, with a typical range of 30–50 metres.
- **Curtain.** Devices which provide cover over a narrow (2 metre) but flat (almost two-dimensional) area. These are ideal for long runs inside the windows of an open-plan office, for example, or mounted sideways to detect entry through a roof or ceiling, and have a typical range of up to 40 metres.

Until recently, PIRs were prone to *'masking'*, which is to say that, by covering the detector, one could prevent it 'seeing' the heat source (intruder). Because it is a 'passive' device, the control and monitoring equipment was unable to detect this masking. However, anti-masking PIRs are now readily available that produce an indication at the control panel when the sensor has been masked. PIRs should not be used in areas which house heat sources or are prone to sudden changes in temperature. Although suitable for most situations, it is advisable to use PIRs in conjunction with other forms of detection in high-security situations, since they are more easily defeated than microwave or ultrasonic detectors. *When they are installed in a suitable environment*, modern PIRs have a reasonable detection capability and high resistance to false alarms. The sensors should be installed as far away as possible from the most likely points of entry and should be positioned so that an intruder's initial movements cross the optical zones at right angles.

Microwave detectors

These devices transmit microwaves (extremely high-frequency radio waves) into the space to be protected. The energy reflects off most surfaces back into the receiver. Moving objects within the area will cause a small change in the frequency of the reflected waves, known as the Doppler effect. This change is detected by the receiver and analysed. Microwave detectors are very effective but, unless they are installed in a suitable environment, they are prone to false alarms. Stable conditions, such as those found in strongrooms, vaults, etc are essential for reliable operation. Moving objects, particularly metal objects like fans or water moving through pipes, are common causes of false alarms. It should be noted that microwaves will penetrate glass and thin partitions. In some cases, this might be an advantage; in others, it may prove a source of false alarms.

Microwave detectors are most sensitive to movements towards and away from the sensors, although in practice the radiation pattern is likely to be so complex that movement in any direction will be detected.

Ultrasonic detectors

Similar in operation to the microwave detectors but operating on high-frequency sound waves, ultrasonic detectors are also very effective in covering large areas at reasonable cost. The ultrasonic waves do not materially penetrate glass or partitions and are therefore suitable for quiet office environments. They *are* prone to false alarms caused by moving air (draughts) and external ultrasonic sources such as disc brakes on vehicles. Ultrasonic detectors are also most sensitive to movements directly towards or away from the sensor, so they should be mounted facing the most likely point of entry.

Dual technology devices

Dual-techs (as they are known) were designed specifically to reduce the incident rate of false alarms, and most modern versions achieve this objective. *Although more expensive than single element devices, the dramatic reduction in false alarm rates is often sufficient justification for the additional cost.* A dual-tech is a combination of two detectors in one housing. Usually, but not exclusively, they consist of a PIR detector together with either an ultrasonic or a microwave unit. In each case, the principle of operation is the same: both detectors must activate within a given period to trigger an alarm. It follows, therefore, that the detection capability of a dual-tech is no greater than the single least effective element within it. For this reason, where a PIR is one of the elements, it is important to ensure that the device possesses an anti-masking feature.

With all three types of movement detectors, it is important to ensure that there are no obstructions between the sensor and the protected area and that the detector 'window' is kept free of dust and grime. The sensors should also be mounted on firm surfaces, as any movement of the sensor will be interpreted by the electronics as moving objects in the protected area.

Additional Electronic Aids

We have now discussed devices used to detect intruders at the perimeter, the fabric of the building and within the building itself. For those businesses whose assets require the physical protection of a safe or strongroom, some additional electronic aids may be necessary.

Light detectors

On the basis that the area within a safe or strongroom is completely dark, it is reasonable to assume that the contents are difficult to remove without the assistance of some form of visible light. A light sensor is simply a photosensitive cell which forms a part of an electrical circuit. Light landing on the cell changes the circuit's characteristics, thereby signalling a change in conditions. Light sensors in totally enclosed areas have two major advantages over other devices, which are:

   they are almost **false alarm free**;
   they provide an **early warning** in the event of penetration by thermic lance or oxyacetylene torch.

One must ensure, of course, that internal lights are switched off prior to securing the strongroom or vault.

Pressure differential sensors

These sensors detect the slight change in air pressure that takes place when a sealed unit is opened. In large enclosures, it may be necessary to employ a fan or blower to maintain the necessary air pressure. Such devices are now rare and have largely been replaced by other types of sensors.

Acoustic detectors

These consist of passive listening devices (microphones) which sample the ambient noise levels and react to sounds above a pre-set level or pattern. Although prone to false alarms unless they are installed in a quiet environment, they have the potential advantage of enabling those monitoring the system to eavesdrop into the protected area, thereby assisting with alarm discrimination. This capability is reviewed later in this chapter when we examine alarm verification techniques (see p 62).

Vibration sensors, heat sensors, volumetric devices and capacitance detectors can all be used to enhance the detection capability of the alarm system protecting a safe or strongroom. One additional point worth making is that it is strongly recommended that the door is equipped with sensors (usually internal magnetic reed switches) that ensure that the alarm system cannot be set unless the door is both shut **and** locked. All your defensive efforts to protect what might be an extremely valuable stock of drugs, diamonds, gold, large amounts of cash or other commodities will have been in vain if the door to the safe or strongroom has been left closed but unlocked (either deliberately or accidentally), because the villains will ignore the alarm system and rely on speed to effect a successful robbery.

CONTROL AND MONITORING EQUIPMENT

Having examined some of the risk-end devices and sensors which should react to intruders, we now need to examine the control and monitoring facilities associated with these. They may vary from a simple box on the wall which, on detecting a change of state at the detector, triggers some form of signalling device, to a desktop computer which displays site plans, pinpoints activated devices and offers reactive instructions to the operator (although this level of sophistication can usually only be justified on manned sites).

All control units need to fulfil several prime functions:

   They must be capable of continuously monitoring all connected risk-end detectors and reacting to signals from these detectors. In turn, this requires a resilience to power failures (battery or generator backup), a resistance to tampering and a high degree of reliability (minimum technical failure).

   They must be capable of activating some form of signalling device (from a local bell on the wall, to a high-security signal to an alarm company's monitoring station).

   They must be able to display the current state of the system, eg set, unset, part set, in alarm, equipment or power failure, etc. In most cases, it is also desirable that the control unit displays which area (or detector) has caused the alarm.

   They must be able to discriminate between authorised and unauthorised users, eg using keys, codes, ID cards, etc.

The control unit itself must be in a protected area. This usually means employing time-controlled access procedures or shunt switches when setting or unsetting the system.

Although not always essential, a good control unit will also provide some additional facilities:

An ability to record (and store for future retrieval) a large number of prior events, such as who changed what and when, all alarm activations, power failures, etc. This is particularly relevant when investigating false alarms.

A facility to permit authorised operators **to add, delete or reconfigure circuits**.

A self-test feature (although this will not guarantee that passive devices are operating correctly).

Where events are signalled externally, it is an advantage if the unit can **transmit detailed information** on individual events and detectors to the monitoring station. Alternatively, it should be capable of processing such events locally to provide point ID alarm verification (see p 63).

Alarm systems are only as good as the last time they were tested and found to be working satisfactorily. This raises the hoary question – should walk-test indicators on detectors be left operative after initial installation? Broadly the arguments run:

For: As the majority of PIR detectors (and dual-techs containing a PIR element) currently in use can be masked (particularly during 'non-set' periods) and this fact is not recognised by the control panel, it is possible to create a situation whereby a silent hour's intrusion would not be detected.

Against: Because the indicators would remain active, this would enable potential intruders (whether staff, visitors or contractors who are present during working hours) to test for areas **not** covered by the detectors.

On balance, it is best to leave the indicators active, for two reasons:

If properly installed, the detectors should not leave any critical areas unprotected.

The indicators facilitate frequent walk-test procedures, without which there can be little confidence that the system remains fully operational. However, walk-tests only prove the system back to the control unit – they do not test the complete system, and in particular the signalling element, which is what we examine next.

SIGNALLING SYSTEMS

The primary purpose of the signalling system is to prompt a rapid response to an alarm activation, when so instructed by the control unit. The characteristics of the more common signalling systems are summarised in the sections that follow.

Local Bells

These are usually mounted on the wall of the premises, sometimes with strobe lights attached. Their purpose is twofold. First and foremost they are intended to frighten off the intruders; secondly, they attract the attention of someone who may react to the situation (preferably by calling the police). The enclosure housing the bell, siren or lights should be mounted in an inaccessible spot, and should be resilient against tampering or removal. Unfortunately, when such alarms are used in conjunction with remote signalling, a conflict of interest exists, as the police will often require a delay of ten minutes before the audible or visual alarm activates; the theory behind this is that the delay enhances their chances of catching intruders red-handed. Unfortunately, ten minutes is often more than enough time for a burglar to complete his attack and escape prior to the alarm being raised – which is not what the occupier (or insurance company) intended. External bells are normally programmed to stop sounding after 20 minutes to comply with legislation on noise pollution.

Remote Signalling

Telephone lines are the most common medium over which alarm signals are transmitted, followed by radio signals. The increasing use of fibre-optic networks (currently used mainly for cable television) opens up an interesting option for the future.

Digital communicator (or digicom)

This superseded the 999 auto-dialler, which is no longer accepted by the police. On instructions from the control unit this device will dial the monitoring station (or any other specified location) over the Public Switched Telephone Network (PSTN) and transmit the relevant data. It is a low-cost device, but suffers from the disadvantage of being unmonitored, so that if the equipment or line fails – deliberately or accidentally – the monitoring station will remain ignorant of this fact and continue to assume that everything is in order.

Direct line (dedicated)

This is the most secure, but also the most expensive, system available. Direct line signalling provides a continuous two-way communications path to the monitoring station, thereby providing contemporaneous information on the state of the alarm system and line integrity, whilst enabling the monitoring station to initiate actions within the protected area.

Telecom RedCare

This is a cheaper alternative to the direct line and provides most, but not all, of the same facilities. A unit located in the protected area is regularly scanned by BT equipment which in turn passes the data on to an alarm company central station over the PSTN.

Radio networks

The most common **radio** signalling system is called 'Paknet'. Like RedCare, this offers two-way data transmissions, but without the need for a telephone network. Paknet is not as secure as a direct line or a RedCare system, but it is resilient to failure and is superior to digicoms. It is ideal as a backup system for high-security installations.

This brings us neatly on to the fourth element in our intruder alarm strategy – the response.

RESPONSE PROCEDURES

It should be obvious that if the response to an alarm activation is inadequate (either because it was too little, too late or was non-existent), then all our efforts and expense will have been wasted. Speed is the key element of an effective response, because the time taken for a reaction force to arrive at the scene of the intrusion must be shorter than the time needed by the intruder to penetrate all the physical barriers. Whenever possible, and certainly in high-security situations, the primary reaction force should be the **POLICE**. As false alarms in the industry continue at more than 90 per cent, police confidence in the integrity and reliability of the system is paramount. To this end a new 'Unified Intruder Alarm Policy' was issued by the Association of Chief Police Officers (ACPO) in May 1995, the aim of which is:

> .....to enable the police to provide an effective response to genuine intruder alarm activations thereby leading to the arrest of offenders and a reduction of losses by improving the effectiveness of alarm systems and reducing the number of false calls to the police.

The policy document covers a number of wide-ranging issues, but its prime purpose is to create a set of conditions which will restore police confidence in the efficacy of alarm systems, which in turn should lead to faster responses.

For lower risk sites, or those with a history of false alarms (which perhaps have been blacklisted by the police), an alternative source of response may be appropriate. The most common of these is an alarm company's 'key-holding' service. For a reasonable fee, a security officer will respond to the alarm, evaluate the situation and react appropriately – for example, by calling the police if intruders are, or have been, on the site, or by checking the source of the alarm and resetting the system, if appropriate.

It should be noted that the police will require the customer:

> ....to nominate at least two key-holders, trained to operate the systems, able to attend within 20 minutes, contactable by telephone and with their own transport.

The requirement may be met by a professional key-holding service or by company personnel. In the case of a company's own personnel having the responsibility of acting as a key-holder, it is imperative that they receive professional advice on how to approach and observe the premises when responding to an alarm call-out, particularly after dark. Physical assets can be replaced, dead key-holders cannot!

To ensure that they are talking to the right person when on the telephone, the alarm company control station will normally issue a confidential ID code to customers and their key-holders. Due to the increasing number of customers handled by a single central station, it is not unusual to find that a common code is issued for each system. If requested, most central stations will agree to issue unique codes for each company employee involved in the administration of the system. This is vital for medium- to high-risk sites as villains are becoming increasingly inventive at subverting alarm systems and the responses. Just a few of their better known techniques are:

- **The 'smash and grab' method.** The objective here is to get in and out, ignoring any alarms, before the responding parties arrive. Good physical security is the best defence to defeat the smash-and-grab merchants.
- **The 'cry wolf' system.** The thieves will deliberately trigger the alarm either to test the response time in preparation for a future operation, or to convince the police and key-holders alike that there is a technical fault and therefore the system should be ignored or switched off.

Disrupting the telephone network to a large area by tampering with exposed British Telecom (BT) junction points, on the assumption that line faults will not be 'policed', or that police resources will be overstretched and fail to respond, and/or that BT will take several hours to restore the lines.

Time the robbery/burglary for specific times like 8–9 am or 5–6 pm which coincide with the periods when most false alarms occur and when, as a consequence, central stations and police are fully occupied.

Company employees acting as key-holders should be mindful of the above tactics when responding or reacting to an alarm condition, as police resources rarely permit more than a cursory inspection of the building.

Visual Verification Techniques

As previously indicated, false alarms are the bane of the intruder alarm industry. Systems that seek to reduce false alarms by confirming that an intruder is present have been slow to come on the market, but some interesting technologies are now available. These include:

Visual verification. Witnessing the scene before, during or after an alarm activation is probably the most reliable method of discriminating between real and false alarms. One of the first technologies to transmit security-related video images over telephone lines was commonly known as SLOWSCAN. This required surveillance of the alarmed area by a standard CCTV camera.

Analogue video information from the camera is transmitted over a telephone line to a monitoring station. The system is usually configured so that a telephone link is established following an alarm activation. Clarity of picture varies with line quality and the video frames also have a relatively slow refresh rate. Both these elements (speed and quality) can be improved by digitising and compressing the video information before transmission. Although this tends to increase costs, the risks may well warrant the investment.

The development of the technology that enables analogue video information to be digitised and transmitted over telephone networks has opened the door to many new and novel ideas in visual verification. One such system (TVX) takes small cameras mounted directly on to a microprocessor and incorporates these within standard volumetric detectors. When the detector activates, a series of 'still pictures' are transmitted over the telephone network to a monitoring station. It remains to be seen if pictures recorded in digital format on disks or tape will be accepted as evidence in a court of law, but even if they are not, there are other significant benefits to both customer and the police.

An impressive 'real-time' demonstration, recently witnessed, showed how cameras in another European country could be monitored from the UK using the Integrated Services Digital Network (ISDN). Not only were the pictures of surprisingly good quality, but the remote control over Pan–Tilt–Zoom and other facilities was remarkable. This opens up almost limitless possibilities for management when considering remote monitoring of the company's premises.

Audio verification. This is achieved by placing microphones within the protected areas to pick up normal sounds and transit them over the telephone system to the monitoring station. This communications link is normally initiated by the intruder alarm system and enables the personnel monitoring the system to make a better informed judgement of the situation prior to triggering a response.

One point should be borne in mind for both visual and audio verification systems. Unless specific safeguards are designed into the system, the technology will permit monitoring station personnel to 'dial up' video or audio (or both) facilities at will. In some cases this may be desirable, but in others it constitutes a major risk to security, not to mention raising concerns over the invasion of privacy.

Point ID verification. Unlike the visual and audio techniques, point ID does not have to involve the transmission of data to the central station for human interpretation, although this is often the case. The decisions can be made locally by electronic logic circuits in the control equipment. Either way, the theory of operation is the same. Provided the alarm system consists of multiple detectors (preferably layered and of different technologies), then an intruder **should** have to activate more than one in order to reach the target. A single activation is therefore likely to be a false alarm. Again, provided that sufficient devices of the appropriate type have been installed, then intelligent processing of the order, interval and frequency of the activations should enable false alarms to be drastically reduced, without losing any material detection capability.

It is estimated that around 35 per cent of all false alarms are caused by end users failing to follow correct procedures when setting and unsetting the system. A number of procedures (some enforced by the ACPO policy) have been introduced by the alarm companies in an attempt to mitigate this problem. Similarly, as line faults contribute substantially to the false alarm total, police will no longer respond to line faults that occur during unset periods.

INTRUDER DETECTION SYSTEMS: SUMMARY

Electronic alarm systems are inexorably replacing people as a means of defending an organisation's assets, particularly unoccupied buildings and sites. Specifying the correct requirements and then selecting the appropriate equipment, supplier and installer to meet those requirements are critical factors, if the twin objectives of early detection and speedy reaction are to be achieved. Because of the widespread problems experienced over many years with high rates of false alarms that result from faulty equipment, bad installation or inadequately trained staff, there is a strong likelihood that the police will cease to respond to installations with a bad history. This in turn is likely to have implications for insurance cover and premium rates, reinforcing the need for businesses to 'get it right' in the search for optimum physical protection.

For those with little or no experience of the security industry, it would be prudent to install equipment which meets relevant British Standards, the most common of which are:

BS 4737 – standard alarm installations.
BS 7042 – high-security alarm installations.
BS 6799 – wire-free alarm installations.
BS 6707 – DIY alarm installations.

If the requirement is to have the system monitored by an alarm company's central station and hence to invoke a police response, then you should ensure that the company tendering for your business is registered with the county police force. Such companies must meet and maintain high standards of service, whilst the quality of their installations will be inspected regularly by an independent inspectorate like NACOSS or SSAIB (see Chapter 13).

ACCESS CONTROL SYSTEMS

Intruder detection systems are primarily designed to protect empty premises, but unfortunately theft is not confined to out-of-hours activity. Walk-in thefts, and theft by employees is an ever-present threat. Automatic access control is a technology which has evolved to mitigate this risk, and forms the subject of the rest of this chapter.

As we saw in Chapter 2, minimising opportunity by limiting access on a 'need to be there' principle may be a key requirement of our defensive strategy. If so, one must consider which form of control best meets our objectives. One of the simplest forms of access control can be achieved by fitting a lock to each appropriate door and then issuing a key to authorised personnel. However, the effectiveness of this system is inversely proportional to the numbers involved (lost keys, personnel leaving, etc), whilst lock replacement (if the keys are compromised) is an expensive exercise. A low-security alternative might be the installation of mechanical code locks. These have the twin advantages of dispensing with the need for keys and facilitating code changes as often as deemed necessary. They are, however, prone to 'shoulder surfing' (a potential criminal observing the user pressing the buttons) and staff communicating the codes to unauthorised personnel. If used on perimeter doors, an additional mortise deadlock is strongly recommended for use during non-working hours.

Manned control points are probably the most effective, if also the most expensive, method of controlling access. The door (or other physical barrier) is controlled by a security guard or receptionist who has the capability of distinguishing between authorised and unauthorised individuals. In high-risk areas, it may be necessary to provide the person controlling access with some degree of physical protection. In situations where multiple access points are required, the costs may be reduced by using a central control point and employing audio and visual aids at the point of entry to assist with identification. Even this method of control rapidly becomes expensive when a large number of controlled access points are involved, particularly if these are required on a continuous 24-hour basis. A popular solution to this problem is *electronic access control.*

ELECTRONIC ACCESS CONTROL

Popularly known as AACS (Automatic Access Control System), an electronic access control system comprises three basic elements:

1. Something the user **has** (card or token); something the user **knows** (code or PIN) or something the user **is** (biometric, unique to the individual).
2. A device at the access point which will recognise one or all of the above.
3. A central processor or computer which, as part of an overall configuration, can process the information from these devices and respond (normally by releasing a physical barrier) according to a set of pre-programmed conditions.

Smaller systems may combine elements 2 and 3. Electronic access control systems offer many advantages over manned systems. These include:

   Prompt blocking of lost or stolen cards. It matters not if the card, key or number is recovered.

■ **Personnel tracking**. A record of an individual's movements through the control points can be maintained, and this facility can sometimes be extended to time and attendance recording for payroll purposes.

■ **Grouping or suiting**. Similar to the master/sub-master system of key suites, but much more flexible.

■ **Hands-free operation**. Using radio frequency (RF) techniques, the card-holder can be granted access without having to present his card to the reader, eg the card (or token) may be carried in a breast-pocket or clipped to a belt.

■ It may form **part of an integrated management system**, if a host computer is used, monitoring and controlling functions such as fire alarms, energy management and intruder detection systems.

Reverting now to the three basic elements of electronic access control systems described above, we shall examine each in turn for their key elements.

Cards (Something we have)

There are many different card encoding technologies on the market today. Table 4.1 compares the relative merits of nine such card types.

Table 4.1 *Relative merits of access card technologies*

Card type	Durability	Security	Cost	Information capability
Wiegand	High	High	Medium	Low
Magnetic stripe	Low	Medium	Low	Medium
Proximity	High	High	High	Low
Barium ferrite	Low	Medium	Medium	Medium
Infra-red	Medium	Medium	High	Medium
Bar code	Low	Low	Low	Low
Smart card	High	High	Very high	High
Hollerith	Low	Low	Low	Low
Inductive	High	Medium	Medium	Low

For high security or specialised applications, two or more of the above technologies may be incorporated into one card.

Three of the more commonly used card types, Wiegand, magnetic stripe and proximity are here described in more detail:

Wiegand. Named after the inventor, it uses a series of magnetically coded, small-diameter wires embedded in the card which, when passed through a reader head, generates a series of binary code pulses. This type of card is difficult to copy and has proved to be reliable. It is moderately priced and has a long useful life.

Magnetic stripe. The introduction of high-density, high-coercivity magnetic stripe technology has reduced the possibility of accidental alteration or erasure. Although inexpensive, the cards are subject to wear and tear and therefore have a relatively short life-span.

Proximity. Based upon RF technology, these cards permit the reader to verify the inbuilt code without the necessity of 'swiping' or insertion. Operating ranges vary with circuit and aerial design, but are usually 12 inches or less. Although expensive to produce, wear is not a consideration as there is no contact between card and reader. They are quite secure and facilitate 'hands-free' operation.

Personal Identification Numbers (PINs – something we know)

Most people are familiar with the concept of *PINs*, since they encounter them regularly through their use of cash cards and credit cards. Some Automatic Access Control Systems (AACS) automatically align PINs with specific cards – the cards themselves are identified by an administration number – so the card-holder must retain the original PIN for as long as the card is in use. More modern systems permit the holder to change the PIN, either when the card is used for the first time or at intervals thereafter. With the latter method, there is a danger that the PIN-holder will use numbers that are easy to remember (*1234, 9999,* etc) or will keep the same PIN for all of his or her personal cards, as well as the AACS. The AACS central computer will store these PINs and may not be as secure as the banks' computers. The security *message* is, if your access control systems do not have a secure method for storing users' PINs, then it is vital that the same PIN is not used for the AACS and the users' cash and credit cards.

Biometrics (Something we are)

The term 'biometrics' is used to describe the technique of measuring a physical characteristic or personal trait as a means of verifying a person's identity. Largely because of the high costs involved, its use is normally limited to high-security applications. Biometric techniques include fingerprint, retinal, hand-geometry, voice pattern and signature matching.

Where processing time is critical, the designer must bear in mind that biometric devices can often take up to 30 seconds to accept or reject an individual, depending on which device is selected.

CARD READERS

The type of card selected will determine the card-reader technology, but here we will limit our examination to three of the most common types of reader.

Swipe. When designed for Wiegand cards, these readers are epoxy filled, making them immune to weather conditions. They operate over a wide range of temperatures and are unaffected by external magnetic fields.

Magnetic stripe readers incorporate a read-head similar to that in a tape recorder, and as such will usually require periodic cleaning. Strong external magnetic fields may affect their reliability. They are also vulnerable to vandalism and in cold, damp climates they may require a heater.

Insertion. Used with Wiegand, barium ferrite, inductive and other cards, this type of reader requires the card to be inserted into a slot. A physically damaged card or misalignment in the slot may cause false rejection problems. Slot readers are vulnerable to vandalism, eg chewing-gum, superglue, etc.

Proximity. Although one of the more expensive options, proximity readers offer the following advantages:

░ speed of operation.
░ vandal resistance – when mounted behind a solid, non-metallic surface.
░ hands-free operation.

For higher security installations, card readers will normally come equipped with an integral digital key-pad. This facilitates the use of personal codes (something we know) in conjunction with the card (something we have).

In addition to criteria governing the selection of a card and the choice of card-reader technology, there are other considerations which the potential purchaser must bear in mind:

░ **Acceptability to users**. Careful measurement of peak flow rates is critical in deciding the number of access points and speed of operation.
░ **Physical hardware** (doors or turnstiles). Space restrictions or aesthetic considerations will often conflict with security requirements. Although significantly more expensive, full-height turnstiles are

usually much better than doors in providing controlled access, as they present a continuous barrier and discourage tailgating. However, they tend to require a fair amount of floor space and are not always aesthetically pleasing.

If doors are used, ensure that **good quality door closers and locks** are fitted.

Ensure that fitting the lock and electric keep does not weaken **the overall structure**.

Ensure that the system is capable of sufficient **expansion** to meet future needs.

SYSTEM CONFIGURATIONS

Having considered the various card and card-reader options, we shall now examine some of the basic system configurations available.

Stand-Alone System

Usually the card reader will control access through a single door or turnstile. It will simply recognise a valid card (and/or code) and release the locking mechanism. This is useful only when small numbers of people, possibly using common entry codes, are involved.

Stand-Alone Intelligent System

An enhanced version of the above, this will enable the use of individually coded cards, card-blocking (for lost or stolen cards), and a limited memory capability for recording movements and unsuccessful entry attempts. It will also permit the use of time zones. Card readers require individual programming, and again the number of users is limited.

On-Line

This offers all the facility of the stand-alone intelligent system, but employs a central processing unit (CPU) to control a large number of card readers. Although this system offers increased flexibility and scope, if the CPU fails for whatever reason, the individual readers will degrade to stand-alone units, and all transaction recording facilities will be forfeited.

Distributed Processing

This is similar to the on-line system, except that the card readers are upgraded to intelligent door processing units (DPUs), sometimes by linking them to a local processing unit. The main difference is that the CPU (usually a microcomputer) is used to program the DPUs and acts as a central monitor and recording device. This enables the system to continue to function normally, even if the CPU is out of action. During such a failure, transaction data is stored locally and is down-loaded to the CPU when the system is restored. Such a system offers greater flexibility because of the increase in processing power, and can often be extended to include alarm management, time and attendance recording, and energy management facilities. This increased flexibility is a very important consideration when designing a large system which requires several access and authority levels, time and location restriction, and event recording.

Some of the more modern systems use Windows™ as the host operating system. This provides a familiar graphical user interface which minimises training requirements and provides on-line assistance through the 'help' feature.

VENDOR SELECTION

Having identified the *system* requirements, you are faced with the problem of selecting a supplier. Clearly, the choice will be limited to those vendors who market your chosen devices, but it is still worth considering the following important points:

- **Experience.** How long has the vendor been in the access control business? Will they supply you with a list of current customers so that you can obtain other users' opinions on the quality of product and after-sales service? Are spares easily obtained?
- **Location**. Are they local to your area, or will you be faced with long delays when assistance is required?
- **Engineers.** Are their service engineers competent and trained on the equipment of your choice?
- **Upgrades**. Are they committed to software (and hardware) upgrades compatible to your particular installation, or will you be left with obsolete equipment as new systems are developed?

Finally, if the system is to be effective it must:

- accept *bona fide* personnel;
- reject unauthorised personnel;
- be acceptable to staff.

We have now identified the essential criteria to be considered when purchasing electronic access control equipment and some of the devices and systems available.

ACCESS CONTROL SYSTEMS: SUMMARY

Controlling access to your premises is a vital element in any physical protection system you implement. The choices of both technology and supplier/installer are extremely wide, making it very important that you choose carefully. This chapter has sought to present the choices, together with those available for detecting intruders, in a logical way, so that you might make an informed decision.

The three key points to remember are:

1. Determine which devices and system configurations best suit your requirements and remember to allow for expansion.
2. Avoid the obvious pitfalls and choose your vendor carefully.
3. Train your operating staff and communicate with your employees.

5

CLOSED CIRCUIT TELEVISION

INTRODUCTION

Police crime figures show that in 1994 over 1500 burglaries of business premises occurred daily. In 1995 these figures increased as computers and computer chips became a prime target for burglars of all types, even those more accustomed to targeting domestic premises. In January 1995 the British Retail Consortium published the results of its survey of retail crime, which showed *inter alia* that there were more than 4.8 million clearly witnessed thefts from shops by customers.

As stretched police resources come under ever-increasing demands, businesses will be progressively forced to protect their own assets by implementing more of their own crime prevention measures.

The provision of static or mobile guards is one way of deterring theft or burglary, but this often proves to be a very expensive option, particularly if the requirement is for round-the-clock protection. Fortunately, security technology can now provide a cheaper and more reliable alternative, either on its own or as an aid to security staff. A particularly good example of such technology is Closed Circuit Television (CCTV).

CCTV has received a lot of publicity in recent years, particularly through its use by some popular TV crime series. However, those very pictures beamed into millions of living rooms serve not only to demonstrate how useful this technology can be, but also how a substandard installation can result in pictures which are of such poor quality that they are almost useless as an aid to identifying the perpetrator(s).

In this chapter we shall examine some of the main component parts of a CCTV system and discuss how they can be assembled to meet the requirements of the user in the best way.

Figure 5.1 shows the key elements which go to make up the most basic standard CCTV installation. Later in this chapter we will examine some additional equipment which can enhance significantly the overall effectiveness of a CCTV system.

Figure 5.1 *The key elements of a basic CCTV installation*

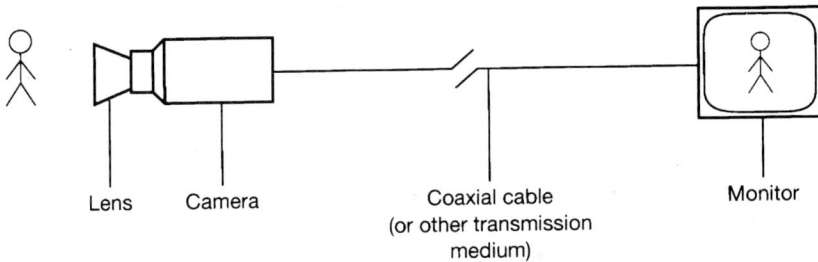

Lens　　Camera　　　　　Coaxial cable　　　　　Monitor
　　　　　　　　　　　　(or other transmission
　　　　　　　　　　　　medium)

LENSES

As those who are familiar with ordinary photography will realise, the lens is one of the most important elements in any image-capturing equipment. Its function is to capture light reflected from the area or object under observation and focus this image on to a light-sensitive surface – in this case the chip or tube of the camera. Already we have determined the main ingredient for an effective CCTV system – *light* – although, as we shall see later, some special types of camera can be effective at very low light levels. Nevertheless, for best results, the target area should be well illuminated and the lens capable of collecting and focusing this light without undue degradation or attenuation.

Lenses are manufactured with certain characteristics that are designed to meet the operational requirements of the user. Some of the more common design features are as follows:

Focal length. This is the distance from the centre of the lens to the plane of the imaging device in the camera and is normally measured in millimetres. The *longer* the focal length, the *narrower* the angle of view, eg a 12.5mm lens will have a much wider viewing angle than a 50mm lens. The *longer* the focal length, the *greater* the magnification.

Aperture. This is the area of the lens through which the light passes. The aperture is normally controlled by the **iris.**

Iris. This controls the amount of light reaching the imaging device and can be adjusted manually or, as is now quite common, automatically to compensate for changing external light levels.

F-number. This is the relationship between the **aperture** and the **focal length** and is expressed as a fraction, eg f1.4; f2.8, etc. The *smaller* the f-number the *greater* the amount of light reaching the camera sensor, and therefore the *'faster'* the lens.

Focus. This permits light from the target to be focused on to the surface of the imaging device in the camera.

The prime factors to be considered when selecting a lens are as follows:

Angle of view. The width and height of the scene you want to observe, and the distance between the camera and the area under observation.

Depth of field. This is the distance between the nearest and furthest points in the target area which can be kept in sharp focus, and is controlled by the *focal length*, the *distance from the camera to the target*, and the *f-number*.

Degree of illumination at the target area.

Mount. Almost all lenses for CCTV use come in 'C' or 'CS' mount formats. *The key thing to remember is that the lens and camera mounting systems must match.*

A wide range of lenses are available for use with CCTV cameras. The most common focal lengths range from 3.5mm (wide angle) to 50mm (narrow angle), with f-numbers between f1.2 and f1.8. Lenses outside these ranges are available but are normally only used in specialist applications.

Zoom lenses are lenses with a variable focal length and, therefore, variable magnification, and should maintain focus on the scene or object throughout the travel of the zoom. In most CCTV installations zoom lenses are motorised and controlled remotely.

Varifocal lenses are fixed lenses which allow the focal length to be adjusted manually.

CLOSED CIRCUIT TV CAMERAS

A CCTV camera is an electronic device which converts light waves into electrical signals, either analogue or digital, depending on the technology used. These are then amplified to a level that is suitable for transmission to a monitor which converts these signals into a visual image on the screen. **CCD** (or Charged Coupled Device) cameras have now largely replaced **tubed** cameras, although the latter are still used for some specialist applications.

There are some important features to look for when comparing CCTV cameras:

Colour or **monochrome** (black and white). The price of colour cameras has come down to the point where, in the context of the overall installation cost, it may seem logical to opt for colour every time. However, as we shall see later when we consider the 'operational requirements', monochrome cameras may still prove to be the most effective option in some circumstances, eg as a general rule-of-thumb, monochrome cameras will provide more useful pictures under very low light-level conditions.

Format. Generally available in ⅓ inch, ½ inch, ⅔ inch and 1 inch formats, the measurements refer to the size of the light sensitive sensor in the camera on to which the image from the target is focused. Lens and camera formats **must** match. The *larger* the format, the *better* the performance, other things being equal.

Sensitivity. Manufacturers usually express sensitivity as 'lux' at a specific f-number (lux being the unit used to measure light levels). The lower the 'lux' number, the more sensitive the camera (for a given f-number). Purchasers should not rely solely on these figures, however, as on their own they do not provide a definitive comparison. *Live tests are still the only satisfactory method of comparing cameras of similar specification.*

Resolution. Normally expressed as a number of lines (380, 450, 580, etc), the *higher* the number, the *sharper* the picture. In low light levels, higher resolution cameras generally give better results.

TRANSMISSION MEDIA

There are a number of ways of transmitting the video signal from the camera to the monitor. First, we shall examine the **terrestrial methods** and then look at some more exotic airborne systems.

Coaxial cable. This is the most common transmission medium. The quality of coaxial cable varies enormously and particular attention should be paid to the robustness and attenuation characteristics. High quality cable should enable video signals to be transmitted over distances of ½ kilometre or more without significant loss of image quality.

Twisted pair. Although the cable itself is less expensive than coaxial, it has the disadvantage of requiring a transmitter and receiver for each video link. However, it can carry the signals over greater distances.

Fibre optics. Fibre-optic cables have two distinct advantages over copper wire systems. They are highly resilient to interference from electromagnetic sources and signal attenuation is very low. This permits high quality signals to be transmitted over long distances.

PSTN/ISDN. The technology which spawned video conferencing over the Public Switched Telephone Network (PSTN) and, more recently, the Integrated Services Digital Network (ISDN), has been adapted to great effect by the manufacturers of security video systems. Surprisingly good quality video (and audio) signals can be retrieved from cameras installed almost anywhere in the world. The pictures are not strictly 'real time' but for most practical security purposes are almost as useful. Like the direct link cable systems, data can be passed both ways, thus enabling the user to control remotely Pan-Tilt-Zoom units, etc.

There will always be occasions when terrestrial systems are either impracticable or uneconomical, in which case there are two **airborne systems** to be considered. Both are *'line-of-sight'* and the transmission distance is therefore limited by the height of the installation (aerials).

Microwave. Viable up to 50 kilometres, microwave links are ideal in or across cities, rivers or lakes where cable links are not feasible. They are reasonably resilient to climatic conditions (particularly the shorter links) and electrical interference. Unless they are protected physically, the dishes (aerials) are susceptible to deliberate tampering or damage.

Infra-red. Only reliable over relatively short distances because of its susceptibility to adverse weather conditions, infra-red has the advantage of requiring no licence to operate. Like the microwave link, infra-red should be considered when cable systems are inappropriate.

Having looked at **lenses, cameras** and **transmission mediums,** the final link in the **basic** CCTV system is the **monitor**.

MONITORS

The monitor is the device which converts the video signal back into a visual image (hopefully an exact replica of what the camera was looking at in the first place!). Like cameras, they come in various shapes and sizes and are manufactured to different specifications.

Colour or **monochrome.** Normally, one would match colour monitors with colour cameras, particularly when monitoring live scenes; however, this may not always provide the best solution. Monochrome monitors, for example, have approximately twice the resolution of a colour monitor and therefore, under certain operating conditions (eg poor light levels), will provide sharper, clearer pictures.

Size. Standard CCTV monitor screen sizes range from 9 inches to 27 inches. These sizes refer to the distance between the diagonally opposite corners of the screen. Smaller screens may be perceived to provide higher definition pictures, but the most important issue when considering size is the distance between the monitor and the observer. For the best results, this distance should be between three and five times the screen size. The area available for mounting the monitors is another consideration governing the size of the screen.

So far we have limited our appraisal to a *basic* CCTV system, consisting of a *lens*, a *camera*, a cable and a *monitor*. However, suppose that we cannot afford the luxury of having the screen constantly monitored by an alert human being, but we would like to be able to review events retrospectively. A *video recorder* will meet our most basic needs, but does that mean changing the tapes every three hours? Alternatively, we may require several cameras but do not want, or need, a separate monitor for each camera nor, for that matter, multiple video cassette recorders (VCRs). Fortunately, technology has advanced to meet most of our needs, but beware, as a general rule, *the more 'in-line' equipment one adds to the basic system the greater the risk of diluting the quality of the observer's view of events.* We shall now examine some of the available peripheral equipment.

VIDEO RECORDERS

The video recorders most commonly used in conjunction with CCTV installations are Time Lapse Video Cassette Recorders (TLVCRs). These enable a single 3-hour tape to last for up to 960 hours. The speed is normally adjustable in steps (3 hours; 12 hours; 24 hours; 48 hours...960 hours). It is important to note, however, that a 3-hour tape running at the 12-hour setting will record only one-quarter of the information, and at the 960-hour setting it will record only 1/320 of the available video frames. This effect is most obvious when reviewing a recording of moving objects. Some will appear to transverse the screen in jerky movements, but with the slower speeds the object may not be captured on tape at all. Almost all recorders in the UK now use the VHS or S-VHS format, the latter giving over a 400-line horizontal resolution. Generally speaking, the more expensive the TLVCR, the more facilities come inbuilt. Some of the more **useful extras** are worth noting:

- **Date/time generator.** Data from an internal clock is recorded simultaneously with the video signals and are displayed on the screen during replay.
- **Audio recording.** Now available on some models at tape speeds up to 24-hour mode, although the quality of the recording will deteriorate progressively as the tape speed reduces from the 3-hour mode.
- **Alarm activation.** An essential feature for most systems, this triggers the recorder into real-time recording speed, thus providing the highest quality picture frames at the most critical periods.
- **Single shot mode.** Enables the tape to be reviewed frame by frame.

■ **Alarm search.** Using the alarm memory, a particular area of the tape can be located quickly.
■ **Auto head cleaning.** The alternative is to use a head-cleaning tape regularly.

At this point it is worth mentioning **video cassette tapes**. These are the standard tapes that we buy for our domestic video recorders. Three-hour tapes are the most common, although four-hour tapes are not hard to find. They also come labelled as 'standard' and 'high'-grade quality. In September 1995 *Which?* magazine published the results of tests conducted on 47 various brands. Surprisingly, they found little difference between the standard and high-grade tapes either in terms of performance or durability. However, experience has taught the author that ***the quality of reproduction reduces rapidly after a tape has been used a number of times***. This is particularly true when the tapes are used with a Time Lapse Video Cassette Recorder. As a general rule, a tape should not be used more than 10 times with a TLVCR or 20 times with a standard video cassette recorder (VCR). However, by regularly reviewing the tapes and recording the number of times they have been used, the maximum reuse rate will soon become apparent.

So far, we have only considered a single camera installation. However, as most systems will require more than one camera, we now need to think about the receiving end. We could simply install a monitor (and a TLVCR if required) for each camera and in some instances that might be the correct solution. However, in addition to keeping costs down, it is usually beneficial to simplify the viewing station or control room ergonomics as much as possible. A room full of monitors and recorders is almost impossible to manage if it is attended and will be surplus to requirements if it is unattended. There are a number of ways of overcoming this.

SEQUENTIAL SWITCHERS

Figure 5.2 demonstrates one method of how inputs from multiple cameras can be configured to terminate on to a single monitor screen. This simple method is commonly known as **sequential switching**. The video signals from each camera are switched, in sequence, onto a single monitor. Additional features include:

■ A *manually* selected, *static* picture on a *separate monitor*, from any of the input cameras.
■ Operator choice of the order of sequence of the scan.
■ The ability to *'skip'* or *'hold'* any camera input.
■ Adjustable dwell time for each input.

■ Alarm inputs which can override manual settings.
■ 'Looping', which allows additional equipment to be connected, eg video tape recorders.

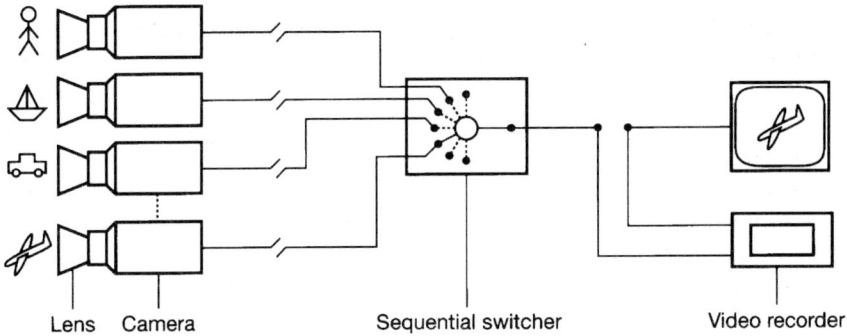

Figure 5.2 *Sequential switching: inputs from multiple cameras terminate on to a single screen*

A major drawback with sequential switching is that the observer (man or machine) can view the output from only one camera at a time. Depending upon the number of cameras (some switchers can handle as many as 64 cameras!) and the dwell time selected, it is entirely possible or even highly probable that significant events occurring in the target areas will go unnoticed and unrecorded. Connecting alarms from the relevant areas to the switcher may mitigate this problem, but not solve it completely, as the cause of the alarm may no longer be in view of the camera by the time the monitor displays the picture. It is advisable therefore to restrict the cameras per monitor (or recorder) to a sensible number (say four), and to keep the dwell time down to a maximum of two seconds. Some TLVRs now incorporate an integral four-camera switcher.

SCREEN SPLITTERS

An alternative arrangement which overcomes some of the above problems is depicted in Figure 5.3.

Commonly called **screen splitters**, these devices enable pictures from several cameras (up to 16) to be displayed on one monitor screen. As the full screen display can be recorded on a single recorder this arrangement has the **advantage** of recording all camera outputs simultaneously. *'Picture-in-picture'* is also possible using digital techniques. As with the sequential switchers, automatic or manual overrides can be utilised to select full screen pictures for any one camera. The more obvious

disadvantages with screen splitting are that the size of the observable picture diminishes as the number of cameras per monitor increases and resolution degrades proportionally.

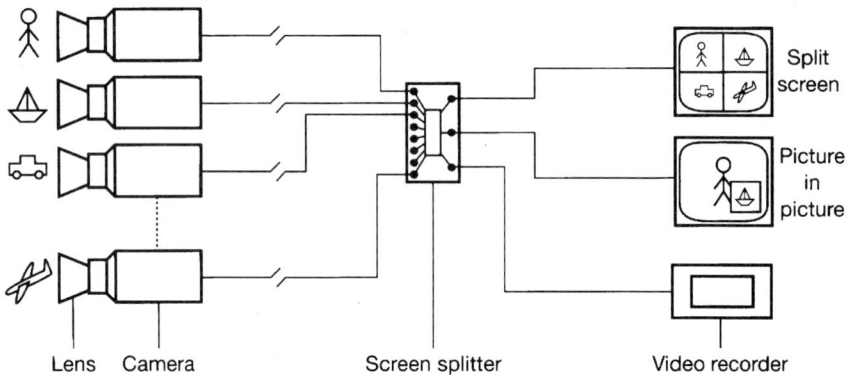

Figure 5.3 *Screen splitters: pictures from several cameras can be displayed on one screen*

VIDEO MULTIPLEXING

Video multiplexers overcome most of the limitations of sequential switchers and may be used with or without screen-splitters. *Video multiplexing* is the name given to the technology which enables the transmission of several video signals virtually simultaneously. A video multiplexer can combine signals from several cameras (usually up to 16) and provide both a multi-camera display on a single monitor and simultaneous recording of all cameras to a single tape recorder. The video signals from each camera are individually coded, which **enables the stored video information from each camera to be recovered and replayed separately**.

The number of frames available for replay will be proportional to the number of cameras connected to the multiplexer (and the recording speed of the TLVCR). One should be careful to ensure that the replay facility is not initiated at the expense of live recording, ie most multiplexers cannot record and replay simultaneously.

As multiplexers are now almost *de rigueur* as an integral element in CCTV installations it is worth explaining (in simple terms) just how they work and some of the system limitations.

Consider 16 cameras feeding into one multiplexer and the output being fed into a time-lapse VCR. The multiplexer will select a frame from camera No.1 and pass it to the TLVCR, followed by a frame from camera No. 2 and so on. As each camera generates 25 frames per second, it becomes obvious that less than two frames per second from each camera is passed through the multiplexer to the TLVCR. Now, if the TLVCR is set to record in realtime (3 hour mode), when you come to replay the tape for any single camera, the picture will look like an old silent movie, only more pronounced, as you will see less than two frames per second. In reality the situation is often much worse than this as there seems little point in buying an expensive time-lapse recorder to run it in the 3 hour mode (a domestic VCR is much cheaper), and in any case one would have to change the tape every three hours. Accordingly, one could expect to find the TLVCR set on the 24 hour mode or even higher. When set on the 24 hour mode the TLVCR records only 1/8 of the video information being passed to it by the multiplexer. Under these setup conditions (16 cameras and the TLVCR in the 24 hour mode) only one frame every four seconds from each camera is being recorded – far from ideal – and of course higher settings for the TLVCR (e.g. 72 hour mode to cover weekends) makes matters even worse, recording only one frame per camera every 12 seconds!

However, all is not lost as modern technology has come up with some clever widgets to overcome some (but not all) of these problems.

Some multiplexers come equipped with a feature commonly called an activity monitor. This is a simple form of video motion detection (VMD) and is used to increase the rate of frame capture from those cameras detecting motion at the expense of those which do not. If, for example, motion was detected on only one camera (the other fifteen viewing unchanging scenes) then the frame capture rate for that camera would increase eight-fold.

The advent of the 24 hour real-time VCR has also helped greatly. By using a clever piece of technology the real-time recorder can capture about three times more frames than a standard TLVCR set to the 24 hour mode.

If the above two pieces of equipment are combined with a reduction in the number of cameras connected to each multiplexer, say to eight, then it is possible to record approximately two frames per second from all cameras under static viewing conditions and up to 16 frames per second from cameras detecting motion within their field of view.

PERIPHERAL EQUIPMENT

We have now considered some of the key components of a CCTV system. Peripheral equipment which may be required for some installations includes:

> **Loop framestores.** These use solid state memories to store continuously a number of frames from each camera. An alarm input will cause these images to be down-loaded on to another storage device

from which *'before'* and *'after'* the event pictures can be recovered. This facility is very effective in discriminating between real and nuisance alarms and is often used in conjunction with video motion detectors.

Video printers. Modern printers can produce high quality hard copy images. These are very useful for longer-term storage and can be very effective when circulating descriptions of suspects.

Pan-Tilt-Zoom units. A *'PTZ'* unit consists of a number of electrical motors, remotely controlled, which will pivot the camera both horizontally and vertically, and enable the user to operate any zoom lens fitted to the camera. Most modern units enable a number of pre-set positions to be programmed into the unit. Upon receipt of an input signal (say, from an alarm covering a specific area) the unit will automatically pan/tilt/zoom to provide pictures from that area. Other useful features include variable pan/tilt speeds, auto-return (after a pre-set period the unit will return the camera to its original viewing position) and auto-pan (the unit will continuously pan between two pre-set stop points).

Environmental housings. All outdoor and some indoor cameras will require a housing of some sort to protect the equipment from hostile elements. Mounted outdoors, a heating element within the enclosure will usually be necessary and a wiper/washer accessory may also be required. *Indoor cameras* are often mounted within mirrored domes which prevent those under observation from identifying the direction in which the camera is pointed.

Control equipment. Modern switcher units will enable the operator to control all the available functions from a single consul, eg PTZ adjustments, multi-camera display changes, recording instructions and remote switching using telemetry. The control panels tend to take the form of either a computer-like keyboard with a joystick, or a touch sensitive screen displaying icons, arrows etc. Each has features to commend it but selection is usually down to operator preference.

Video motion detectors. Potentially one of the most useful and cost-effective elements of any CCTV system, video motion detectors (VMDs) have failed to live up to the high expectations of users. There are two principal reasons for this:

— Significant technical improvements to the early versions have been rather slow to emerge.
— Users have consistently installed VMD equipment in systems and environments that are patently unsuitable to its capabilities, thus contributing to its reputation as a major source of false alarms.

Nevertheless, as VMD equipment is installed in series with the video signal at the control end and can be fitted retrospectively, it is often a very cost-effective method of enhancing a CCTV system. Video motion detection works by comparing, frame by frame, the grey scale levels at individual points within the video picture. Thus, a person or object moving within the target area will create discernible changes in the video signal which can be detected and used to attract the operator's attention or trigger a recorder. Unfortunately, these changes in the video signal may also be caused by camera movement, rain or snow, headlights of passing vehicles and many other natural or man-made phenomena. Algorithms have been developed which help to distinguish between wanted and unwanted signals, by comparing the speed and direction of motion, and the size and position of the object within the image, but have met with varying degrees of success.

Some systems utilise several cameras to create a '3-D' zone, whilst others combine the VMD with another intruder detection system. Both these methods are effective in reducing the incidence of false alarms. Technology is now gathering pace and ever more intelligent algorithms are emerging which will ensure that video motion detection systems will become an integral part of a CCTV system for outdoor as well as indoor use.

KEY CONSIDERATIONS

We have now considered most of the hardware which is available to us, but how much of it do we need? Some very valuable work is being carried out at the Home Office Police Scientific Development Branch (PSDB), where staff are seeking to establish a performance criteria, based on customer requirements, which can be tested on an on-going basis and readers are advised to study the relevant PSDB publications carefully before making large investments in a CCTV system. Here, we will look at some of the more significant issues to be considered.

Assuming that an effective risk assessment has been carried out and that a decision has been made to install CCTV as part of a loss-prevention programme, the performance objectives need to be clearly identified. CCTV installations can usefully be considered in two modes – **action** and **information** systems:

Action systems are those where the video information will be used contemporaneously to make decisions or to take action. A simple example of an **action** system would be where a camera is installed at the entrance to an area or building to which admittance is restricted to those persons or vehicles which are recognised by the person monitoring the screen. Based upon this visual information access is granted or denied.

Information systems are those which require no immediate action but will monitor and record video images, for days, weeks or even months, in order to provide a visual record of events. Such information may be vital to the success of any subsequent investigation.

One significant difference between the two modes is that any **deterioration in the performance** of *action* cameras, because they are in continuous use, tend to be noticed and remedied, whereas a similar trend in an *information* system may not be spotted until a review of the tapes is called for, by which time it may be much too late. Although *all* types of CCTV installations should be subject to regular maintenance and performance testing against predetermined standards, this is especially important with regard to information systems.

However, when designing a CCTV system, there are some common operational requirements to be considered:

- Is the objective to **detect** a change in the picture, such as an object or person moving within the field of view; or will there be a need to **recognise** the object, eg is it a small or a large dog, a male or a female person, or a particular make of car; or will there be a requirement to **identify** the object, eg distinguish between two members of staff? The objective will determine the size of the image required for display on the monitor screen, from the former (10 per cent of screen height) to the latter (100 per cent of screen height).
- **How large is the area to be covered?** Bearing in mind the above objectives, how many cameras will be required, remembering that the depth-of-field diminishes as the angle of the lens increases?
- Will the system be required to operate during the **hours of darkness?** If so, are the lighting levels adequate? (See p 85 on lighting.)
- Will a **moving target** need to be tracked by the system?
- Will the video images need to be **recorded and stored?** If so, for how long? Does the recording need to be continuous or can it be triggered by an event detector?
- Will the observer need to be **alerted to events** within the target area?

Careful consideration of the above requirements is essential if the unnecessary purchase of unsuitable equipment is to be avoided. One simple technique which may help a purchaser to keep things in perspective is continually to ask the question, *'Is this particular piece of equipment* ***essential, desirable***, *or just* ***nice to have?'*** The essential parts are easily identified, the desirable bits need to be carefully analysed in terms of cost-effectiveness and the nice-to-have components can rarely be justified (although quite often when various pieces of equipment are compared, all of which meet the user's requirements, one finds 'extras' on some at no additional cost).

LIGHTING FOR CCTV

One of the most important considerations for an effective CCTV installation is the provision of adequate lighting. Although camera manufacturers will claim varying degrees of sensitivity, eg $\geq$ 0.1 lux, this will not necessarily guarantee good quality pictures at this level of illumination. The following table may be used as a rough guide, but again it is stressed that there is no substitute for live testing.

Lux	Description
50,000	Bright summer day
5,000	Cloudy day
500	Brightly lit office
15	Main road lighting
5	Side street lighting
0.3	Full moon
0.001	Clear sky starlight

Some cameras are designed especially for low light-level operation. These incorporate **image intensifiers**. Simply put, image intensifiers are light amplifiers, so it is immediately apparent that *some* light is necessary for them to operate satisfactorily. Another technology which is becoming ever more relevant to CCTV is **thermal imaging**. These cameras employ front-end devices which are sensitive to radiation emitted by hot bodies or objects. Such cameras will not provide, of course, anything like the quality of image that a normal camera, operating in good lighting conditions, produces; but, as always, it is a matter of '*horses for courses*'.

There are several key issues to be addressed in addition to the intensity of the illumination:

   The type of lighting is important. A good way to demonstrate this is to take your camcorder out into the street at night and record for a few minutes. On replay you will notice that the colour and contrast are both very poor. This is because normal street lighting comes from **low pressure sodium** lamps which have a very narrow frequency spectrum. **High pressure sodium lights** are better and **metal halides** are even better. The broader the frequency spectrum, the better the colour and contrast, all other things being equal.

   Lighting the target area. There is no doubt that lighting the target area locally is the best, if sometimes the most expensive, option. Fitting spot- or flood-lights alongside the camera may seem an economical solution, but it produces satisfactory results rarely. The exception might be where the requirement specifies that the target needs to be tracked over unlit territory. In such a case, the light should be switched off when it is not required.

   Covert lighting. Most CCTV cameras are sensitive into the **infra-red** region of the frequency spectrum, some more so than others. As human eyesight cannot operate at these wavelengths, it is possible to observe events within a seemingly 'dark' area by illuminating the area using infra-red lighting (usually white light with an infra-red filter). This may also be the only solution in situations where white light would cause problems – for example, near domestic dwellings or public highways.

COMMON PROBLEMS

Having looked at the criteria for an effective CCTV installation, it may now be prudent to review some of the more common problems which lead to dissatisfaction among users.

- **Spot- or flood-lights mounted alongside the camera.** In heavy rain or snow, the reflected light will obscure the target area. In thick fog, it is most likely that no usable images will be captured (this is especially true of infra-red lamps). Filament bulbs used for infra-red illumination tend to fail after about 2000 hours of operation and are quite expensive to replace.
- **Pan-Tilt-Zoom units left in the wrong position**. This is a serious problem when the camera associated with the PTZ plays an integral part in the overall coverage pattern. An **auto-homing** facility (where the PTZ returns to a pre-set position 'n' minutes after it was last used) is strongly recommended.
- **Unstable camera mounting.** In windy conditions, the vibrations will seriously degrade picture quality and totally negate any associated VMD systems.
- **Control-room ergonomics.** Cramming the monitoring station with monitors will simply make the operator's task more difficult. A clear distinction between **action** and **information** systems, together with 'attention grabbing' signals and regular recycling of monitoring staff, should help to ensure that the original objectives are achieved.
- **Unrecorded (or poor quality) images.** Usually these are due to human error in failing to replace tapes, clean the recording heads, or faulty programming of the recorder. Employing the *'KISS'* principle (keep it simple, stupid!), do ensure that your requirements will be met by using only one tape per machine per day. A 'library' of 31 tapes (numbered 1–31) will provide for a fresh tape each day (tape no 1 on the first, no 14 on the fourteenth, etc). Not only will this method be easy to administer on a daily basis, it will provide one month's record and the tapes can all be changed simultaneously after one or two years' usage (depending on the type of recorder installed). Too many cameras connected to a single recorder can lead to images being missed altogether.
- **Image problems.** Maladjusted focus accounts for a large proportion of poor quality images, which are particularly apparent when

reviewing tapes. This is most common on remotely controlled systems as the focus adjustment is often coarse and delayed. Poor light and trying to cover too large an area with a single camera are the other main causes of poor quality images.

EMERGING TECHNOLOGY

There are currently a number of interesting developments that hold out the prospects of worthwhile advantages over existing technology. These include:

- **Digital storage.** This refers to the technique of converting *analogue* video signals into *digital* data, thereby offering the possibility of storing this data on hard disks or DATs (Digital Audio Tapes). Some companies are already offering this facility, but it is expensive. The longevity and quality of the stored images, plus the speed of search made possible by the technology are its prime advantages.
- **Flat screens.** There is now a genuine chance that *affordable* flat screens will be available within two years. Their advent will revolutionise control room ergonomics.
- **Computer control** of CCTV systems, using *Windows*-based software or touch-screen technology, provides a high degree of control flexibility (eg auto-patrolling) whilst enhancing user-friendliness.
- **Advanced colour CCTV cameras** are now available which automatically revert to monochrome operation in poor lighting.
- **Video compression and transmission techniques** have now advanced to the point where one can monitor, and control, CCTV systems remotely over standard telephone lines (even from the comfort of your home using a laptop and a modem).
- **Digital cameras** are probably the cameras of the future although the bandwidth necessary for transmission of full resolution uncompressed video is beyond current PSTN or ISDN telephone lines.

SUMMARY

CCTV, properly specified and intelligently installed, can play a significant role in the management of risk. User expectations are most likely to be met where an accurate assessment of the risks is undertaken first, a realistic view is formed of the part that CCTV can play in reducing those risks, equipment is selected that can achieve the objectives and an on-going review is made of how the equipment is performing. Hopefully, the information in this chapter has identified key aspects in the choice and use of CCTV to help combat crime, so that any investment in this technology is much more likely to be justifiable after the event in terms of its contribution to the achievement of corporate objectives.

6

SECURITY IN THE OFFICE
Fraud, Embezzlement and the Loss of Information

INTRODUCTION

Like it or not, modern business is riddled with dishonesty, fraud, deceit and corruption which, too frequently, take managers by surprise. It has been said that for every credibility gap there is a gullibility fill, and this is unarguably true. People simply don't ask the questions they should, often in the belief that to do so is impolite. (Comer, Ardis and Price, 1988)

Those are the words of Mike Comer, founder of the very successful investigation company, Network Security Management Ltd.

The Mori survey referred to in the *Sunday Times* article, published in June 1995, seven years on from Mike Comer's observations, provides yet another endorsement of the parlous state in which we seek to conduct our business:

More than two-thirds of Britain's leading finance directors say their companies have been victims of serious fraud by employees, a survey reveals today. The report discovered an atmosphere of diminishing job security where managers, using inside knowledge of company security, lined their own pockets. Other executives are falsifying accounts to make their departments meet increasingly demanding performance targets.

The really worrying aspect of these two comments lies in the different perceptions of the role of managers. In 1988, managers were 'too frequently taken by surprise' by fraud and dishonesty, whereas by 1995 it is the managers themselves who are leading the way. This represents a significant culture shift, with serious implications for security.

In this chapter, we shall look at some of the vulnerabilities associated with office work, of which fraud is just one. As before, we shall also consider some of the available ways to counter the problems. Office activities are usually at the hub of any business, and our review will include accounts and finance, purchasing and information, a broad spectrum within which lies much that is crucial to corporate survival.

WHAT IS AT RISK?

As generally with security, any review of vulnerabilities in the office benefits from starting with a risk analysis. *What is it in my organisation that is at risk? From what or whom is it at risk? What are the likely consequences if the risk materialises?* It is only by answering these questions that appropriate countermeasures can be devised, measures that also relate the cost of security to the scale of potential harm.

So what might be at risk in an office? Top of most people's list are likely to be the **accounting processes**, embracing purchases and sales, payroll, expenses, petty cash, banking, and so on. Certainly these represent substantial risks, particularly when you consider that the fraudster's greatest need is to convert the proceeds of his crime into money or goods. However, **procurement fraud** has recently emerged as a very real threat and is often associated with **entrapment and bribery**. In addition, organisations increasingly rely for their competitive edge on the exploitation of **privileged information**, derived variously from their own research, from market intelligence, from unique production formulae and processes, from knowledge of their customers and suppliers, from the evolution of business strategy, from capital investment plans and cash-flow forecasts. Any loss of such information – either in absolute terms, such as might result from a major fire, or in terms of their unique ownership of it, such as would occur if the information was made available to a competitor – would almost certainly prejudice the organisation's performance and might, in extreme circumstances, lead ultimately to its decline and downfall.

If information has only recently assumed such a pivotal role in corporate affairs, **embezzlement** was first defined in law during the reign of George III (1760–1820), to cover the offence of larceny by a servant or clerk. An illustration that the problem has not gone away is provided by some research (Grey and Anderson-Ryan, 1994), within an American national chain of restaurants:

> We have found that all employees who have worked in the restaurant more than 1 year scam, or about 60% of workers. Newer workers tend to scam less because, "they just haven't figured it out yet – just give 'em time".

This case study, incidentally, also illustrates the importance of corporate culture on crime and criminality. A culture that openly encourages and acknowledges ethical and honest behaviour will exert a strong positive influence on employees, certainly, but also on suppliers and customers.

Their perceptions are likely to result in reduced risks of collusive crimes, of bribery and of entrapment.

We shall look in detail at these areas of risk later, but the second part of our risk analysis is concerned with identifying *what* or *who* are the sources of the threats, and it is this aspect that we shall consider next.

WHAT OR WHO POSES THE THREAT?

There is a danger in what follows of appearing paranoid (you remember the old adage, *'Just because I'm paranoid, it doesn't mean the bastards aren't out to get me!'*), since the list of potential threats is a very long one. Some of these are fairly obvious, but there are others whose propensity for inflicting harm on your organisation is much less apparent.

The principal threat must be from the **employees.** Consider these facts and predictions:

- in the UK more than one in three men and one in eight women will be convicted of a standard list (criminal, non-motoring) offence;
- the average embezzlement lasts 42 months before it is discovered;
- polygraph tests on American bank and retail shop employees showed that between 40 and 70 per cent had stolen from their employer;
- a survey by the London School of Economics indicated that nine out of ten schoolboys admitted stealing by the time they left school.

(Comer, 1985, p 5)

As the *Sunday Times* report (1995) indicated, with regard to the problem of fraud, 'employees' should no longer be taken to mean those working at a lower level than the manager. Rather, employees should be taken to include all members of the workforce, from senior directors down to the lowly manual worker:

> *Fraud comes down to the people you employ, and how you check on them and manage them. How well you spot if someone is in trouble, with serious financial or personal problems.*

> *The risk/reward balance has changed for employees and people now feel easier about taking the risk.* (Deville and Jenner, 1995, p 17)

In the case of fraud, especially, success usually depends upon the exploitation of a long-standing relationship between the perpetrator and the victim – a factor that clearly places the employee in a privileged position. Thus, clerks, supervisors and managers working in accounts, purchasing, sales and marketing departments, are all likely to have ready access to monies and goods, and therefore represent a prime potential risk, whilst there are likely to be many other employees whose work provides them with the occasional opportunity to steal or defraud.

After employees, the list of fraudsters is a long one. An emerging area of risk, but one where the full impact has yet to be documented, is that of **contract workers**. The recent move to 'delayer', 'down-size' or 'right-size' organisations has led to significantly fewer permanent employees, particularly supervisors and managers. They have been replaced, as and when fluctuating corporate workloads demand, by outsourced workers, hired on finite term contracts. The principle of outsourcing is long established, but used to be restricted to 'temporary' typists and clerks hired in to cover for holidays and sicknesses; now it is applied wherever there is a need for temporary help, and includes those with scarce technical skills, project management skills, and professional expertise of all kinds. The practice begs a number of questions. How thorough are the pre-employment checks for these contract workers? Can they be expected to develop similar loyalties to the organisation to those employed on more permanent terms? What level of supervision should they be given? Can they be expected to respect company confidential information after they have left one temporary employment and moved on to another? Is their next employer likely to be a direct competitor of the current one? As can be seen, there are several new areas of risk associated with the practice, which have yet to be fully addressed by those responsible for the changes in corporate structure that have produced the situation.

'Contractor' is also used to cover those various trades people who, from time to time, visit your premises to build, repair, alter, paint, service machinery, and so on. These have long represented one particular kind of threat, best summarised in the definition, *'A contractor is someone whose prime task in life is to burn down or otherwise destroy your property'*. Apart from the threat to physical safety and security they often represent, contractors, of course, do have access to and contact with staff, some of whom will have the power to award or withhold business, and others who will authorise and arrange payment for the service provided. They are therefore potential partners in collusive fraud and the possible source of bribes and backhanders.

Competitors are a threat, especially over the security of corporate information. A great deal of information can be obtained about a company from the public domain: annual reports, public relations statements, patent registrations, job vacancy advertisements, market and sales intelligence, and the like. Sometimes, however, this is not enough, particularly during mergers and acquisitions, or when major strategic decisions are being formulated. At such times there is a heightened risk of industrial espionage through surveillance, through bribing or suborning an employee, or through placement of an employee within the target organisation. Sometimes, **careless talk** by employees in public places (during a flight is a favourite, but also in hotels, restaurants and bars) is responsible for the release of sensitive information; and any competitor knowing of this possibility would make it his job to discover the opposing MD's favourite eating places.

Customers and suppliers are sometimes liable to collude with your own staff in frauds against the organisation. Common examples are bribing a sales ledger clerk or supervisor to write off the legitimate debt due for goods or services, in return for a 'commission'; or, on the supply side, bribing a purchasing manager to favour the particular supplier involved. We shall look at this problem later.

Investigative journalists represent a very real threat whenever they scent 'a story'. This might concern the rumour of an impending large order, especially if this is itself a sensitive issue, such as one that derives from a country with a bad 'image' in the UK. Other issues likely to attract journalists and unwelcome publicity might concern environmental pollution, a serious accident, financial problems, directors' remunerations, a possible takeover, staff redundancies, legal writs and boardroom wrangles. Clearly, many of these could well impact negatively on the business if you are unable to control the way in which the news is released, or if any release into the public domain is liable to cause problems.

Beware also of the potential for harm from **sales and technical representatives,** whose propensity for social chat and natural inquisitiveness can produce situations where company-sensitive information is revealed to them inadvertently by the employees they visit. There is then the distinct possibility that that information will be resurrected and thrown into the conversation during subsequent calls they make, so that before long what was confidential information has become common knowledge. The worst scenario is one where the representative releases the information to a competitor. **Consultants** represent a similar source of potential indiscretion, even though the nature of the consultant/client relationship should be based on absolute trust and confidentiality.

Last in the list of people-oriented threats, though not necessarily the least significant, are **criminals** – usually in the guise of vandals and burglars, sometimes as robbers, and occasionally as arsonists. In any of these manifestations, they can cause considerable financial loss and disruption; in the more violent cases, the impact of their assault upon your premises and/or staff can be traumatic and long-lasting.

So far, we have considered only the impact and intervention of people, but **accidents and natural disasters** also have the potential to cause great harm to your business. Fires and floods represent the most frequently occurring natural disasters – just think of the impact of the recent floods in Chichester, which virtually closed off the centre of the town for many days, whilst any large town near you will have experienced several fires within the last few years which will have severely affected many local businesses. On a smaller but even more common scale, **accidents, errors and omissions** during routine office processes are recognised as the source of significant business losses, something that emphasises the importance of sound systems and good controls.

This has been a broad review of the risks associated with office activities. It is now time to examine in more detail how threats manifest themselves, before ending with an examination of the measures available to counter them.

HOW AND WHERE THE THREATS OCCUR

In General

It is worth making the point that 'white collar crime' is usually conducted by first-time offenders – that is to say, offenders who have not been caught before. The most persuasive theory why people engage in fraud and embezzlement is that propounded by the criminologist, Edwin Sutherland, who explained it in the following way:

> The data which are at hand suggest that white collar crime has its genesis in the same general process as other criminal behaviour, namely **differential association.** The hypothesis of differential association is that criminal behaviour is learned in association with those who define such behaviour favourably and in isolation from those who define it unfavourably. A person in an appropriate situation engages in such criminal behaviour if and only if the weight of the favourable definitions exceeds the weight of the unfavourable definitions. (Sutherland, 1961)

It is arguable from this that an organisation or its directors which condones or deals weakly with crime actually encourages others to join in. Crime is contagious. By contrast, Sutherland's theory implies that an organisation which shows a strong sense of right and wrong, which encourages honesty and is ethical in the way in which it conducts its affairs is thereby ensuring that 'the weight of the definitions unfavourable to crime' exceed those favourable to it.

Another criminologist, Donald Cressey, helped to explain why it is so often the trusted employee who commits fraud:

> Trusted persons become trust violators when they conceive of themselves as having a financial problem which is non-shareable, are aware that this problem can be secretly resolved by violation of the position of financial trust, and are able to (rationalise their behaviour to themselves). (Cressey, 1973)

Finally, Emile Durkheim's theory of *anomie* suggested that as long as a person's aspirations are balanced by the opportunities available for achievement, a state of contentment would exist. However, should these aspirations be incapable of fulfilment through legitimate opportunities, then the person so thwarted would turn to unconventional methods instead – one of which could be crime. Furthermore, a recent study (Sarah Willott, 1998, reported in *The Independent*, 11 September 1998) suggests that lawyers, accountants and other middle-class professionals (many of whom ran their own firms) who were convicted of fraud refused to accept they had done anything wrong, but believed instead that they were morally superior to 'common criminals', and argued that they were stealing to provide money for their families, or to keep their businesses and staff afloat. One convict said, 'Although my family's grown up now, at the time this happened I had a young family. Then you look to your staff, who in turn have got responsibilities and young families themselves. My crime was taking money, not for my personal benefit, but for the benefit of others, to keep the firm going.' Willott concluded that upper-middle-class offenders were able to retain the moral high ground, despite entering the alien working-class underworld of prison.

The lesson for employers here lies in the need to pay particular attention to the aspirations of very able and intelligent staff, for some of whom the opportunities for career progression are likely to have been reduced, rather than improved, by the recent trend towards delayering organisations.

Within Accounts

Fraud and embezzlement are the prime risks in accounts departments, whilst the greatest number of threats emanate from the victim organisation's own employees. Those frauds reported most frequently concern:

- the creation of fictitious invoices from often non-existent suppliers, leading eventually to the issue of a cheque in payment from the victim company to an account controlled by the person committing the fraud. A variation of this fraud is the inflation of invoices from a supplier, in which the overpayments that result are shared between supplier and the victim company's employee;
- the issue of false credits (or sometimes the failure to invoice at all) to a favoured customer, which reduce the amount legitimately owed to the company and the consequent underpayment of a debt. Usually this is undertaken in return for a bribe from the customer, but can be the result of a family relationship or even entrapment of the employee by the customer, leading to extortion;
- the creation of fictitious employees on the payroll, resulting in regular payments to accounts and/or addresses controlled by the perpetrator. Concealment in this case is often achieved by introducing a high number of leavers and new employees, so that there is less chance of the fictitious names being discovered by others;

- the manipulation of stock records (now most commonly held on computers) to cover genuine thefts of goods and materials from the stores;
- the theft of monies received as payments from customers, by manipulating the records in such a way that the consequent shortfall is covered by falsely crediting other receipts from different customers (known as teeming and lading, or lapping frauds). This is typical of a 'drip type' fraud, where it usually starts with small amounts, but rapidly escalates into a complex morass of many dozens, if not hundreds of false entries having to be made every month, merely to keep the fraud rolling and to hinder discovery. One fraudster in Nottingham some years ago eventually gave himself up to the police and confessed all, because, he said, he was having to work a great deal harder to keep the fraud moving than ever he did before he started it! Beware, in this context, the trusted employee who never takes holidays. Is their constant presence required to keep a fraud from falling apart?
- The theft from petty cash of regular, small amounts, which are concealed by failing to record the associated transactions.

Before moving on to examine problems in other office-based activities, let me ask you what you believe would be the maximum loss a ten-person organisation could suffer from petty cash: £1000? £5000? More? I am going to give you the bare outlines of a petty cash fraud that occurred within a subsidiary of a multinational company, lasted for at least six years undetected and grossed *at least* £300,000!

The monies were extracted by the bookkeeper by misappropriating cheques drawn to fund petty cash on no less than 137 occasions. Typically, fraudulent payments would occur twice a month for amounts between £1500 and £3000 each, and the sums were charged initially to the cost of sales, but were then changed to a VAT recoverable account. Despite the long-running nature of the fraud, the external auditors consistently signed off the annual accounts without any qualification, claiming later that the books totalled arithmetically and that their random spot checks 'failed to hit these fraudulent withdrawals'. The most recent internal audit took place three years before the fraud was discovered.

The bookkeeper had 22 years of service and gave no evidence of an extravagant lifestyle or of the need to sustain an expensive habit like gambling or drug-taking. At one point during the fraud, the bookkeeper had such an overdraft on his current bank account that the bank cancelled his credit card. He currently maintains that he has no property or any of the stolen money, whilst the police have confirmed that he has no significant assets in this country.

An internal report commented:

> The fraud highlights the well known risks of operating with small business units which cannot support enough staff for adequate segregation of duties.

Control is therefore dependent on the supervision of the general manager and his integrity.

At low levels, fraud may go undetected for several years and can thereby accumulate to not insignificant sums. Measures to eliminate the risk in small units would not be cost effective however.

Within Purchasing

Some idea of how seriously fraud in general is now viewed can be gleaned from the regularity with which the subject is discussed in broadsheet newspapers, on radio and television and in journals which are not overtly concerned with security or accounting. As a case in point, the following extract is taken from *Director,* the house journal of the Institute of Directors, drawing attention to a recent growth in one type of fraud:

Alongside the breakdown of traditional feelings of corporate loyalty has come the latest buzzword of 'empowerment' – allowing autonomy within the management structure and devolving authority down the chain. For fraudsters this is a gift; for honest employees, perhaps battling to keep their jobs and match expectations, it is extreme temptation.

*Partly as a result of all this, **procurement/entrapment fraud** has emerged as a real threat.*

All the consultants interviewed for this article confirmed that procurement fraud is fast becoming the main problem – meaning anything from kickbacks, bid-fixing and appropriating company assets to all-out entrapment and extortion. With one individual in charge of (say) purchasing, the potential for fraud is enormous. (Director, 1994, pp 44–48)

The article then gave examples, which included the acceptance of payments by the Norwich Union's print purchasing controller in return for awarding printing contracts to one particular printer, and another where tendering details and other confidential information were sold by employees in a civil engineering company, resulting in one supplier being able to raise its bid some £20 million above what it would otherwise have submitted.

It has long been recognised that those who place orders with outside suppliers have the potential to exploit this power for personal gain. Stories abound within the construction industry of house extensions, garden landscaping, new garages, and so on being provided as an inducement (sometimes the words used are 'expression of thanks') for the award of contracts. Less clear-cut is the present of a case of wine or whisky, a hospitality day at Cheltenham races, the Lords test match or Twickenham, or a night at the theatre with dinner after. A good rule-of-thumb is that all gifts received have to be declared to one's immediate superior and 'pooled' within the department. A slightly different approach is to refuse any gift unless you are in a position to reciprocate. Without some such safeguards and house rules, the risk of purchasing managers laying themselves open to entrapment by unscrupulous suppliers must remain high.

Remember, too, that it is not just those located within the Purchasing Department who decide what shall be bought and from where. Many managers have the authority to specify particular goods or services, including those responsible for security, IT, production, transport, catering and communications (telephone systems and faxes). It is often the case, especially with technology, that the specification itself severely restricts the possible sources of supply, which in turn places the specifying manager in a similarly privileged, and potentially manipulable, position to those exclusively concerned with purchasing.

International recognition of the seriousness of procurement fraud has recently come with an initiative from the 29 industrialised countries of the Organisation for Economic Co-Operation and Development.They produced the Convention on Combating Corruption of Foreign Officials in International Business Transactions at a meeting in Paris in December 1997. This is designed to make bribing of foreign politicians or officials a criminal offence, and is to be backed by anti-corruption laws in the signatory countries. It would end incentives such as tax breaks for bribes paid abroad, which are allowed for French, Swiss, Spanish and German companies.

The aim is to curb the standard practice of secret bribes and commissions paid in countries such as the Gulf States, India, Indonesia, Kenya, Mexico, Nigeria, Russia and Zaire. The UK government intends to put a memorandum ratifying the convention before Parliament in the autumn of 1998, and altogether seven countries have draft bills before their legislatures to bring the convention into force. These include Germany, Japan and Belgium; whilst the USA has already approved the convention in July 1998. (They already had in place the 1977 Foreign Corrupt Practices Act, which banned bribing foreign officials.)

It is clear that the extreme competition for business within our global economy has eroded the historical perception of what is right, what is acceptable as legitimate business practice. What is also certain is that the traditional expression of British disdain for the 'foreign' practice of awarding hefty commissions in return for the placement of business can no longer be justified by reference to the allegedly superior business ethics that apply here.

Within Information Technology

There is a separate chapter (Chapter 8) in this book about computer security, but it is relevant here to make one or two general points about risk in this area. The adoption of computers and networks has fundamentally reshaped the way in which we work. Before this technology was available, we all made a number of assumptions and relied upon what Professor Piper calls 'social mechanisms' for security. Thus, we assumed the integrity of the information contained within a sealed envelope, we authenticated the origin of the information by recognising the signature at its foot, and we adhered to the concept of discussing it in 'secret' behind closed doors.

However, Piper says, these assumptions are undermined once we use computer systems to conduct our business:

> ...and the consequence can be disastrous. Furthermore, frauds may occur in transaction systems by people impersonating legitimate users and gaining improper access, or by hackers gaining access to critical data which is held on-line... If, for instance, two people are communicating by e-mail, then they cannot see each other and it is not immediately obvious how either of them can establish the identity of the other. Furthermore, unless precautions are taken, anyone storing information on a database may not be able to control who has access to that information.
>
> Similarly, anyone receiving a message over a network will need to ask themselves the following:
>
> Am I confident I know the identity of the sender?
> Am I happy that the message I have received is identical to the one which the originator sent?
> Am I concerned that the originator may later deny sending this message and/or claim to have sent a different one? (Piper, 1995, pp 192–194)

Most computer users are those with a PC on their desk, and most of these are connected within networks. One of the main advantages claimed for this technology is that it improves communication and speeds access to information. Even in 1993 it was reported that over 60 per cent of large companies allowed external trading partners access to their computer systems – concepts like Electronic Data Interchange (EDI) are based upon the *business* advantages of providing such facilities – and this proportion will have risen in the years since. Unfortunately, the search for convenience often runs diametrically opposed to the provision of security. It is one of the most important tasks of senior management to balance the advantages of the technology against an informed assessment of the risks, before deciding how much of an overhead in the form of security to build into the system.

A prime example of the dilemma posed by weighing commercial advantage against increased risk can be found in electronic trading through the Internet. There is a developing consensus that such trading will rapidly become very important, if not vital, by providing a world-wide market place for goods and services. However, a 1998 report by *Price WaterhouseCoopers* and *Information Week* magazine revealed that the recent growth in electronic commerce, enterprise resource planning (ERP) and supply chain management software have increased corporate security risk. Significantly, 59 per cent of companies conducting business through their web site or implementing electronic supply chains and ERP applications reported at least one security breach during the previous year. Worryingly, nearly half the 1,600 companies surveyed could not even tell if they had lost revenue as a result of security breaches (report in *The Financial Times*, 1 September 1998).

CONTROLS AND COUNTERMEASURES

Any plan to counter the various threats both to and bound up with office-based activities will need to be broad in its scope. Countermeasures will include aspects of:

- Physical security.
- Corporate culture.
- Personnel security.
- Information security.
- Internal audit.
- Computer security.
- Management controls, including preventive financial controls.
- Fraud detection/proactive fraud investigation measures.

Physical security and computer security are big enough topics to warrant chapters of their own in this book, so they will not receive more than a passing reference here. The others will be discussed in the remainder of this chapter.

Physical Security

Without reasonable physical security, many of the other available methods of combating office-based crimes will be seriously undermined. Without the means to keep unwanted visitors out of the premises, for example, it would be extremely difficult to manage an effective information security programme or to prevent thefts of personal computers. So attention to access control, secure windows and doors and to means of detecting any breach of physical security must come high on everyone's list of security priorities.

Physical security should also include an awareness and assessment of the environment in a broader sense. **Crime Prevention Through Environmental Design (CPTED)** is a strategy based upon the ability to reduce the level and fear of crime by proper design and effective use of the built environment (see Chapter 3). It is a concept that makes use of environmental psychology, but which is firmly grounded in common sense and practicalities. For an excellent guide, see Timothy Crowe (1991).

Although separate from the CPTED concept, physical security should also consider those aspects of the environment which are capable of causing major disruption. Thus, it would be wrong to locate the computer system in a basement if the building is on a river bank, or to fail to install some kind of impact-resistant barrier outside the building if it is situated on the outside of a bend in the road at the bottom of a steep hill.

Corporate Culture

We have referred in this chapter to much expert opinion that stresses how important the influence of corporate culture is on attitudes to crime and the propensity to criminal behaviour among employees, customers and suppliers. There are widely shared concerns that current organisational trends involving delayering and the increasing use of short-term contract workers negatively influence employee loyalty and honesty, whilst this already difficult situation is likely to be exacerbated by the intensifying competition for business, because dubious and illegal practices (namely, procurement fraud) are increasing.

In such a climate, security risks are clearly increased. Conversely, managers have considerable potential to reverse this trend by the influence they bring to bear on the culture of their own organisations. They can create a climate of honesty and ethical behaviour by personal example; by issuing policy statements on the business ethics they expect to see applied by their staff and also on security; they can rationalise the issue of employment 'perks'; and they can confront the particularly difficult issue of employment tenure openly and fairly with those of their staff who are likely to be affected. There has been much research to support the validity of *equity theory*, which holds that employees respond very positively, even under conditions of stress, provided they believe that they are being treated fairly. It has been found to apply to working conditions, work loading and levels of pay, and represents another opportunity for managers to influence staff behaviour.

Personnel Security

One management control that experience indicates is not used as rigorously as it might be is that exercised over new employees by means of **pre-employment checks**. Security staff sometimes use the word 'vetting' in this context, but this term has unfortunate overtones. However, it must be sensible that, before you invite a stranger to join your payroll and become privy to the way you do business, you at least check their educational and employment history.

Security case law is long on examples of employees who have been hired despite being less than fully honest about themselves. The kinds of things for which an employer should look out include:

- Overstating educational qualifications.
- Unexplained gaps in the continuity of a candidate's employment. *(Was the candidate temporarily detained 'at Her Majesty's pleasure'?)*
- Overstating the seniority and responsibility of previous job(s).
- Reasons for leaving previous employment. *(Was it just a 'difference of view' with the boss, or something more serious?)*
- Failure to state any criminal convictions. *(Remember, 36 per cent of males aged 30 in the UK have been convicted of a serious criminal (non-motoring) offence.)*
- Failure to state qualifications markedly in excess of what the job requires. *(What are the motives? Is the candidate likely to be able to exploit those over-qualifications for criminal gain – for example, a cleaner with a degree in computer science?)*

References should always be taken up, but not just the two provided by the candidate. Instead, all employers during the previous ten years or so should be approached for confirmation of the period of employment, the nature of the work undertaken and the reasons for leaving. A very good test question to ask previous employers, because it does not commit them to any statement that might not be sustainable in a court of law is, *'Would you employ this person again?'*

A sensible approach to pre-employment screening is first to rank the jobs for which you are recruiting in terms of the potential harm a miscreant job-holder might cause the organisation. Those jobs which rate highly in this respect warrant more care (and cost) over the pre-employment checks that need to be made. Thus, more sensitive jobs might justify checks into the candidate's credit worthiness, or psychometric tests to help determine important personal characteristics.

We have advocated often enough in this book that security must be the servant of the business, that it must offer cost-effective solutions to problems that restrict the business' ability to achieve its objectives. Appropriately thorough pre-employment checks represent one of the most cost-effective crime-prevention measures any manager can introduce. How much better it is to eliminate potential trouble before it arrives than to endeavour to control and contain it when you actually experience it.

Information Security

Much has been written about information security, but it can be distilled into four principles:

- Devise and implement a simple classification system (the three categories, *'Secret'*, *'Company Confidential'* and *'Unrestricted'* or *'Public'* cover most needs, although some would add *'Personal'* to these).
- Educate staff in the importance of keeping sensitive information secure.
- Establish procedures for handling, processing and storing information that are appropriate to its classification. (This will inevitably include aspects of computer security, but should also embrace secure filing cabinets and a clean desk policy.)
- Devolve responsibility for the classification of the information to its originator or 'owner', who should also be the person who undertakes any subsequent review and possible reclassification.

Even from such a brief outline, it can easily be seen that keeping information secure can become a major task. Like many major tasks, therefore, the advice is to tackle it as you would that of eating an elephant – a little bit at a time! There are two important points to remember: do your risk assessment first, and don't introduce additional security unless it will enhance the performance of the business.

On two levels – that of protecting information about individuals and that of ensuring, as far as possible, that employers only recruit and retain honest employees – data protection legislation in the UK and Europe has a key role. Since 1984, this has depended in the UK on the Data Protection Act: but in 1998 the UK government introduced a new Data Protection Bill, which had its origins in European Commission (EC) directives seeking to harmonise European Union (EU) member states' approaches to safeguarding information obtained and stored about people. The Bill has received the Queen's assent and will become law in January 1999. The new legislation no longer limits its remit to personal information kept on computers, but extends to virtually all data, because it includes:

- information that is being processed by means of equipment operating automatically in response to instructions given for that purpose;
- information that is recorded with the intention that it should be processed by means of such equipment;

and

- information that is recorded as part of a relevant filing system.

The Bill is based upon eight data protection principles, the first of which states that 'Personal information shall be processed *fairly and lawfully*.' These terms are then governed by tests described in Schedules Two and Three, which are too extensive to detail here, but which should be studied and understood by all those responsible for obtaining and processing information – which in practice means just about all managers.

It is the Data Protection Registrar's definition of what constitutes 'fairness' that will cause difficulty for those concerned with security – and particularly the investigation of suspected breaches of security. She has said that 'When you are obtaining information about a data subject from a third person, that third person must be made aware of your identity, the purpose for which you require the information, and the identity of any other persons to whom you intend to pass on the information.' In effect, you will not be able to obtain information by deception, commonly known as pretext enquiries.

Since the first clause of Schedule Three states that the data subject must give their explicit consent to the processing of sensitive personal data, there must also be doubt about the legality of pursuing investigations and processing the resulting data where a crime against the business is being investigated. It would seem that such data might be considered 'sensitive', and if this were so then the data subject's 'explicit consent' would be required beforehand. Can you picture the situation? 'Excuse me, Mr Jones, but I should like your consent to process any information I obtain about you during the course of my investigation into the theft of £5,000 from petty cash.' At the time of writing, there also remains substantial doubt about the future legality of covert surveillance, for this is not possible if one first has to inform the suspect.

Such considerations have serious implications for the investigation of suspected crimes. Whilst there is a clearly justifiable and proper concern to protect fundamental human rights and freedoms, notably the right to privacy, this needs to be balanced by a recognition of the rights of organisations to protect themselves and the assets they control from the actions of criminals.

A related issue is the possibility of prohibiting what is known as 'enforced subject access'. Under Section 21 of the 1984 Data Protection Act, individuals have the right to obtain copies of computerised personal data which are held about them. This is called subject access. Enforced subject access occurs when one person (say, an employer) requires another (say, a prospective employee) to exercise this right (say, by applying to the police for a copy of the police record about them), and then to pass the information on to them in order to prove that they do not have (in this example) a criminal record, before the offer of employment will be made. The 1997 Police Act provides for expansion of the availability of police checks, so that employers should be able to obtain information on an applicant's criminal record, but only about those convictions not yet spent under the Rehabilitation of Offenders Act 1974. However, the processes embodied here threaten to be overly bureaucratic and with inbuilt delays.

It is a common view in official circles that the only important crime is that involving theft and violence. However, this overlooks the serious impact of fraud – not just the major business fraud that occasionally makes headline news, but the vast range of smaller-scale business crime, fraudulent insurance claims by both business and individuals, fraudulent claims for social security, and a plethora of dishonest financial activities, often facilitated by electronic cashless systems. The majority of such crimes are not investigated by the police, but are dealt with by businesses themselves, either through corporate security departments or by hiring private investigators. Restrictive new laws would tie the hands of legitimate investigators, protect the criminal, deny justice to the vulnerable and increase the costs of services to the public. For example, it is generally accepted that insurance premiums are 5 per cent higher than need be because of fraudulent claims.

An example of the practical way in which the law, technology and the investigation of crime are increasingly likely to conflict is revealed in a recent ruling of the Data Protection Registrar. Elizabeth France decreed that local authorities were breaking the law when they demanded disclosure of staff payroll information from employers. Their aim was to match data against benefit claims to spot fraud. The 1997 Social Security Fraud Act allows tax and benefit records held by local authorities and social security and other government departments to be data matched. It also provides powers for council inspectors to investigate where they believe someone has been committing benefit fraud. However, the Data Protection Registrar adjudged that 'It does not give anyone the right to undertake fishing expeditions. Wholesale data matching exercises are a major invasion of the private lives of people to whom no suspicion attaches' (report in *The Financial Times*, 16 July 1998).

One thing is certain in a generally uncertain world, and that is that information will remain the lifeblood on which corporate life depends. Information Technology has created unprecedented opportunities for gathering, collating and manipulating information, and it is right that the ways in which we use this power should be closely monitored and regulated. The current debate is about whether the correct balance is being achieved between the rights of the individual and those of the organisation.

Internal Audit

Internal audit, like security, is not a function universally supported by small- and medium-sized businesses. Nevertheless, if the nature of the business activity produces a serious exposure to the risk of fraud (as it might, for example, in financial services, in long-running project management, or in relatively complex manufacturing processes), even fairly small businesses might benefit from the special skills and insights of a trained auditor.

One important difference between an internal and an external auditor lies in their responsibility for detecting fraud. The external auditor will argue by reference to the standard conditions of audit that it is not part of his duties to investigate the existence of fraud, whereas the internal auditor should be deeply involved with the establishment of financial and accounting controls and adherence to them. Although this concern need not necessarily translate into an active investigation *of* fraud, it should certainly lead to the identification and reporting of **non-compliance** with procedures designed to eliminate the possibility of fraud.

A number of computer packages are available on the market designed to assist the detection and investigation of fraud. Among them are *ACL* and *Idea* (which are good, flexible audit tools), *Net Map* (which plots complex relationships, as might occur in mortgage fraud) and the research tool, *SPSS*, which has been used by auditors for regression analysis.

Idea is PC based, costs around £800, and can be used to capture data in any format for subsequent manipulation and analysis. Thus, purchase invoices can be screened for duplicate payments, for (suspiciously) quick payments, or for break-point clustering, which is the clustering of payments that fall just beneath an authorisation limit. It can be used on outside purchases to analyse, for example, the corporate outgoing telephone calls, to determine if any individual members of staff make multiple calls to particular numbers, which is of potential significance in collusive fraud and purchasing fraud, or even in the loss of information.

SPSS is a particularly useful technique for linking dependent variables within a business – for example, the value of shop sales with the cost of staff wages – to highlight significant variations between (in the example used) different branches. It is important to emphasise that any significant variations outside an established norm will not point unerringly to the existence of fraud or theft, but it will enable the investigator to identify the areas where further examination is warranted. Using such statistical techniques, it is possible to build up quite complex models of the business or particular processes within it (production, transport, sales) so that actual performances are compared to theoretical norms and significant differences highlighted for investigation.

Computer Security

This is a sufficiently large and important subject to justify a chapter to itself, so please refer to Chapter 8 for a reasonably full examination of the risks and countermeasures.

Management Controls

Over 40 per cent of fraud incidents involve losses of £50,000 or more and the trend is upwards. It is therefore only prudent for management to implement controls that make the commissioning of fraud more difficult.

In a landmark case, *Re: City Equitable Fire Insurance Company Limited, 1925*, directors were held to have a duty to act at all times with skill and care, 'such as could reasonably be expected of a reasonably prudent person and the degree of skill which may reasonably be expected from a person of that director's knowledge and experience'. In order for managers to be able to exercise properly their powers with skill and care, they must establish systems and procedures to run the company on a day-to-day basis, safeguard its assets, process accounting and other transactions, and produce financial and other management information. It follows from this that it is primarily management's responsibility to prevent and detect fraud.

The steps that management can take to fulfil these responsibilities for preventing fraud include:

- install and operate an effective accounting system;
- establish and operate an effective system of internal controls;
- ensure employees understand the corporate code of conduct and that sanctions are strictly applied when this code is broken;
- monitor legal requirements and ensure compliance.

Establishing effective **internal controls** is the key. These are defined within the *Auditing Standards and Guidelines* (1990) as:

> ....the whole system of controls, financial and otherwise, established by the management in order to carry on the business of the enterprise in an orderly and efficient manner, ensure adherence to management policies, safeguard the assets, and secure as far as possible the completeness and accuracy of the records.

Note that it is the *whole* system of controls, which will include physical and organisational controls, as well as those concerned with procedures.

Whilst physical security is dealt with elsewhere, it is worth noting here two important aspects of organisational controls. First, it is very important to have in place a detailed plan of the organisation, which defines and allocates responsibilities, lines of reporting and delegation of authority. Secondly, it is difficult to overemphasise the importance (or to overstate the number of times when it is missing) of the general control that separates those responsibilities or duties which would, if combined, enable one individual to record and process a complete transaction. Ideally, the following duties should all be segregated:

- authorisation of business decisions;
- execution of business decisions;
- custody of assets;
- recording of accounting transactions.

Finally, management should ensure that there are proper controls for the authorisation and approval of all transactions, that there are effective arithmetical and accounting controls to ensure the accuracy of records and business processes, and that there is in place a formal and adequate system for the supervision of business transactions.

Fraud Detection and Investigation

Mention has already been made of the availability of computer-based audit packages that can be used to speed up the investigation of fraud or to analyse data to give an indication of the likely presence of fraud. Such packages are now in common use, although those that are run on large mainframe computers (such as *Net Map*) can be costly. In any case, it makes financial sense to concentrate initially on measures which are designed to prevent, or at least severely restrict, the opportunity for fraud, such as those on which we have so far concentrated.

However, some global situations are commonly accepted as good indicators of fraud. These include:

- Unethical behaviour of management.
- Weak, autocratic and/or remote management.
- Management 'by crisis' – high staff turnover, under-staffing, circumvention or avoidance of authorised procedures, lack of forward planning.
- Decentralisation ('empowerment') not accompanied by adequate supervision.
- Businesses undergoing restructuring ('right-sizing'), especially if there are redundancies.
- Complex projects and long-term contracts.
- Remote sites.

- Large numbers of outside suppliers and contractors.
- Uncertainty over the final outcome of projects. (Will they be profitable, subject to litigation?)
- Rapid expansion of the business. (Have controls and systems kept up?)
- Low morale within the Accounts Department.
- Inadequate/incomplete accounting records.
- Accounts containing a large volume of low-value transactions.
- Commission and personal loan accounts.
- Chances of fraud being detected perceived to be low.
- Unduly lavish lifestyle of employees.

Detection of fraud can be achieved through the operation of management controls, by audit testing and by luck or chance. Some examples of controls that will detect fraud are:

- Regular (weekly/monthly) manual review of the payroll by a senior manager, which will detect 'ghosts' and inflated payments.
- Regular reconciliation between the accounting records and the bank statements, which will detect *teeming and lading* frauds.
- Within a retail environment, reconciliation of sales and stocks.
- Thorough reviews of actual performances against budgets. (The £300,000 petty cash fraud mentioned earlier should have thrown up a substantial variation of this kind.)

Some examples of routine audit testing are the periodic stock count and reconciliation to the stock records, and the review of accounting records for unusual items and any individual transactions classified as 'material', that is, sufficiently large to impact upon the business on their own.

Analytical review is also used to detect fraud and is the technique of systematically analysing and comparing related figures, trends, ratios and other information to detect any inconsistencies or unusual events. Some examples of this are:

- Comparing current with previous periods, with budgets and with similar businesses in the same industry.
- Investigating variations from norms or from forecasts, testing possible explanations and obtaining corroborative evidence.
- Computing relevant ratios and trends from both financial and non-financial data. Thus, corroborating financial and, say, production data is particularly useful.

Luck and accident also have their part to play. Disgruntled mistresses have been known to bring 'pillow talk' to the boss's attention, once their former partners have left the love nest! Informants are not infrequently a productive source, sometimes in the shape of junior staff suspicious of a senior colleague, sometimes from customers and suppliers. Some firms have set up 'Information Lines', telephone numbers which informants can call anonymously with their suspicions.

It is also notable how many frauds come to light 'by accident', although, in fact, the *accident* often occurs when the perpetrator of the fraud is absent from the scene of his crime for some reason – holidays or illnesses are the most common causes. When this happens, the fraud-ster's work has to be allocated to a colleague, who then stumbles on something which is not fully understandable, or which is suspicious. This leads to questions and evidence of the crime gradually emerges. It is for this reason that there is merit in insisting that employees do take their annual holidays and are absent for at least one period of seven consecu-tive days. Where employees are active computer users, a good house rule is one that deregisters the user from the computer for at least seven consecutive days.

Finally, there have been several developments that harness the number-crunching power of computers to perform speedily data reviews and comparisons that would simply not have been possible by manual means. Taking just one area where these techniques can be applied – payroll – checks can rapidly be completed on:

- the validity of employees' national insurance numbers. (There is an algorithm for creating these, of which the fraudster might be unaware.)
- duplication of national insurance numbers. (A fraudster might find it difficult to create a valid NI number, so another employee's number is used for the 'ghosts' that have been added.)
- a comparison of employees' surnames for duplications. (It could be innocent, eg a husband and wife both working, but it could be that one of them is paying himself twice.)
- a comparison of bank account numbers to which salaries are paid for possible duplication. (Again it could be innocent or guilty, but it will also highlight the existence of 'ghosts' on the payroll.)
- duplicate home addresses (for similar reasons.)
- a comparison of bank account numbers between payroll and pur-chase ledger suppliers. (What is the nature of the relationship? If a husband works for you and his wife for your supplier, is there an enhanced opportunity for fraud?)
- a comparison of bank account numbers between an organisation's payroll and its creditors' or customers' accounts, to which payments are made by credit transfer. (The risk of fraud would exist if one half of the partnership were able to destroy the record of a debt due to the organisation.)
- the validity of the PAYE tax codes. (A fraudster might not be able to provide a 'ghost' with a relevant code, or to support it by appropriate monthly returns to the Inland Revenue.)

Even from a relatively straightforward application like this, you can see how much cross-referencing and calculating can be undertaken to detect any events which are worthy of further investigation. Similar constructs can be produced for purchases and sales, for raw material usage and finished goods, and for stock/stores items in general.

CONCLUSION

Because office activities usually form the hub of any enterprise, securing the various spokes that connect the business to the larger environment in which it operates, the security risks associated with it are proportionately high. On the other hand, because accounting is an ancient science, the risks are generally well known and have inspired the evolution of a wide range of countermeasures. There is, however, serious current concern with the possible consequences of what is perceived to be a breakdown of the traditional loyalties between employer and employee, caused by the extensive reorganisation of enterprises to meet the challenges of global competition and to exploit dramatic advances in information technology. If this perception is correct, it represents a new kind of threat, for which the remedies have yet to be fully developed and tested. For the moment, the best advice must be to remain aware of the possibility of crime and to reinforce the management and systems controls that have stood the test of time, especially if the controls have lapsed as a consequence of these very same changes in corporate structure and personnel.

References

Auditing Standards and Guidelines. Regulations issued by the Audit Practices Committee of the Consultative Committee of Accountancy Bodies, 1990.
Comer, Michael J (1985) *Corporate Fraud* (2nd edn), McGraw-Hill Book Company, London.
Comer, Michael J, Ardis, Patrick M and Price, David H (1988) *Bad Lies in Business*, McGraw-Hill Book Company, London.
Cressey, Donald R (1973) *Other People's Money*, Patterson–Smith, New Jersey.
Crowe, Timothy D (1991) *Crime Prevention Through Environmental Design*, Butterworth–Heinemann, Stoneham, USA.
Deville, Tim and Jenner, Peter (1995) *Combating Fraud,* PA Consulting Group, London.
Director, 'The manageable risk', by Nigel Page, London, September 1994.

Grey, Mark E and Anderson-Ryan, Wendy 'Serving and Scamming: a qualitative study of employee theft in one chain restaurant', in *Security Journal*, Vol 5, No 4, October 1994.

Piper, Fred 'Information Security', in *Intersec,* Vol 5, Issue 5, Walton-on-Thames, UK, May 1995.

Sunday Times, 'Big business plagued by white-collar fraud', Ian Burrell, Adrian Levy and John Waples (reporting on a Mori survey commissioned by Security Gazette and Control Risks Group), 4 June 1995.

Sutherland, Edwin O (1961) *White Collar Crime*, CBS, New York.

7

EMPLOYEE THEFT

INTRODUCTION

Although we discuss aspects of criminal behaviour by employees in Chapters 6 and 8, these are limited to office activities and to the use of computers. The purpose of this chapter is to describe the many other ways in which dishonest employees can and do exploit their privileged *insider* status to steal from the organisation for which they work.

Some of the examples we shall examine will be well known to many managers, whilst others might not be such obvious risks. The list of problem areas to be considered here include:

- Stores operations.
- Scrap and waste-materials disposal.
- Staff purchases.
- Stealing time.
- Sales vouchers.
- Expenses.
- Exhibition samples.
- Goods received.
- Management theft.
- Running a business within a business.
- Stop and search policies.
- Recruitment procedures.

An obvious question to pose at the outset is, *'How much of a problem is theft by employees?'* It is one that has greatly exercised the minds of criminologists and other researchers, but which has proved difficult to quantify. Most agree that it is employees who perpetrate the majority of crimes committed against an organisation – they have, after all, the opportunity and the technical knowledge of the systems that are to be exploited. Among the more interesting recent pronouncements on the problem are:

Between 50% and 70% of an organisation's losses are due to internal theft.
(National Retail Association of America)

Employee crime amounts to between 1% and 2% of a company's turnover.
(Neil Snyder, Professor of Business Studies at the University of Virginia)

*Part-time employees who steal take 33% more on average than full-timers
– $414.61 compared to $311.18.* (Results of a research study conducted by
Reid Psychological Systems in the USA)

*The property most likely to be taken by employees is that which they con-
sider to be of **uncertain ownership**, such as disposable items that will
eventually be thrown away and scrap materials from the manufacturing
process. Ultimately, dishonesty in the workplace is a management problem, not a law enforcement problem.* (Dr Richard Hollinger of the Center
for the Study of Criminology and the Law at the University of Florida, based
upon earlier research by David Horning)

Such comments reveal a serious attempt to get to grips with a sensitive
issue, and one, moreover, where judgements often founder on different
interpretations of what constitutes permissible behaviour. Some action
that may be acceptable in one organisation is quite likely to be considered unethical or dishonest in another. The salesman's inflated expenses, an employee's private telephone calls, or the practice of reselling
goods purchased with the benefit of a staff discount are just three examples where attitudes are likely to differ between companies.

The purpose of this chapter is to illustrate the nature of employee
'theft' through case studies, even though it is likely that some readers will
wonder at times what all the fuss is about. Many of the examples chosen
have been taken with permission from cases handled by Hoffmann
Investigations, based in Amsterdam. The principal, Gerd Hoffmann, is a
good friend with a wealth of experience, who regularly publishes details
of cases (maintaining, of course, the anonymity of both client and criminal) to promote a wider understanding of the problem. Details are given
in the Reference section at the end of the chapter.

STORES OPERATIONS

This is an area where the problems are well recognised. Whether the
particular stores contain raw materials, work in progress, finished goods,
foodstuffs for the canteen, or engineering maintenance materials, the
contents are likely to have a value for someone, sometime, somewhere.
The following cases illustrate different aspects of the problem.

POOR CONTROLS

Parts used in vehicle repairs were supposed to be booked out by the Stores Manager, but only at the end of the working day. Both the salesmen and mechanics withdrew parts, and investigation revealed that one of the mechanics was also in the habit of booking parts out. Unfortunately, he included parts that he took home for his own use, as well as those he used for the firm. What is more, he admitted to taking other parts that he did not book out at all.

This case reveals very weak or non-existent controls over stores items. Parts should always be booked out by one nominated person, whilst their use should be justified by properly authorised job sheets and requisitions. There is a need periodically to reconcile stock holdings with purchases and usage through a stock count completed by non-stores staff.

COLLUSION WITH SUPPLIERS

One of the duties of the stores clerk was to book goods in and then out as they were used on internal jobs. Frequently, he made both entries, one immediately after the other. Investigation revealed that this was done because the goods were never delivered (even though suppliers' despatch notes were available to support the inward transaction), and the speedy double entry was made to avoid the risk of any other storeman noticing that the goods were not on the appropriate shelf. The suppliers maintained the stores clerk's co-operation in this fraud by allowing him free shopping from their catalogue.

Frauds of this kind should come to light if 'reasonableness' tests are applied to the costs of maintenance work, or to the costs of keeping production machinery working. If raw materials are the items concerned in the store-man's scam, then actual consumptions can be measured against forecasts to identify any deviations. In any case, withdrawals from stores should always be supported by a properly authorised requisition. In this case, the funda-mental control of the segregation of duties should apply, which would mean that the store clerk's signature would not be acceptable on its own as an authority for withdrawing stores items.

STEALING TO ORDER

The scene is the local pub, where one of the regulars is overheard saying that he is about to paint the outside of his house, but wonders how much the undercoat and gloss will cost. His conversation is swiftly, but politely, interrupted by another drinker, who proceeds to offer him an 'exceptional deal' on some bankrupt stock he has just obtained. Once it is discovered that, by happy chance, the would-be vendor has exactly the shades required, prices are agreed and delivery is promised for two days' time.

Such a scenario features regularly in film and television programmes, and the producer has little doubt that every viewer will understand exactly what is going on. The theft that either precedes or follows such encounters need not be restricted to employees working in company stores, but clearly such people are favourably placed to carry it out. Those who steal in this way are seeking to supplement their orthodox earnings and have other potential sales outlets, apart from not-quite-by-chance encounters in a pub. Market stalls, car boot sales, friends and family can all be used to dispose of stolen goods.

The countermeasures available to the manager will include, of course, good, documented controls on all materials purchased for company use, whether or not they pass through the stores. However, the occasional visit to local markets and the Sunday morning car boot sales, concentrating on those traders who are handling similar merchandise to that produced in the manager's factory, is quite likely to reveal the presence of his own goods, if any are there. This will not uncover the theft of paint or timber lengths, but it is a useful check on finished goods.

SCRAP AND WASTE DISPOSAL

Many organisations seem not to realise the intrinsic value of scrap and the potential losses they might suffer if it is casually or inadequately controlled. All manufacturing processes will produce some scrap or waste materials as a by-product. In addition, some manufactured items will be rejected at quality control stage. It is not an uncommon practice for scrap and waste to be put aside into some form of temporary storage, pending its final disposal and removal, or its recycling into the production process.

The problems can occur if adequate protection is not provided by way of secure storage until one or other of these outcomes occurs. Even where the value of the scrap is recognised by the organisation to the extent that it concludes sales and service contracts with merchants to remove it, it is often the case that the scrap itself is not secured on site, because nobody can visualise anyone but the authorised merchant deriving value from it. However, Hollinger identifies scrap materials clearly falling into that category of *'uncertain ownership'* which, he says, proves to be most attractive to dishonest employees. This reinforces the need to protect such items, if their full value is to accrue where it belongs.

SCRAP ENGINEERING COMPONENTS

This case relates to a vehicle component manufacturer and illustrates how matters can soon get completely out of hand when controls break down. The company made a wide range of car parts, which it supplied to many of the major car manufacturers. Quality control was commendably thorough, so that even though manufacturing processes were sound, a significant number of parts were rejected as being below specification. Reject parts were moved off-line into an area designated for scrap and, because the workings were steel and other metals, most were intended ultimately for the on-site foundry, where they were melted down for recycling.

A strong local market developed for cheap car 'spares', which were initially supplied by employees smuggling off-site the appropriate items obtained from the scrapyard. The legal consequences for the firm could well have been dire – for example, they could be sued by a car owner because the disk brakes they made failed, resulting in a crash in which someone was killed. Even in the unlikely event that they were able to prove that the linings were intended for scrap, how difficult it would be to defend themselves against a charge of failing to secure such items from theft.

However, the employee thieves (as is so often the case) became even greedier and looked for ways of increasing their 'profits' from the trade in components. Their 'solution' to the problem of meeting such a strong demand was to bring one or two Quality Controllers into their scam, thereby ensuring a rather higher rate of 'failures' than had been the case. The failed, but in fact perfect components were surreptitiously marked by the Controllers, so that they could be readily identified among the piles of other scrap waiting for the foundry. The whole system rapidly became so refined that orders were taken from customers one day, were passed immediately to the Quality Controllers, and were fulfilled either the same day or the day after.

This case highlights the dangers of weak or non-existent controls over scrap. It also raises another sensitive issue – the question of whether or not to operate a stop-and-search policy as employees and contractors enter and leave company premises. The issues will be examined later in this chapter, but in this particular case such a policy should have prevented the excesses that occurred.

WASTE CONTRACTORS

The cost of removing waste from a site can be very high, especially for items such as chemicals and other products containing toxic materials. Most organisations do not remove such waste themselves, but contract with a specialist waste-disposal firm instead, which is usually paid on the basis of the number of lorry- or container-loads of waste it removes. One does not have to be too cynical to note that it is not unknown for waste contractors to invoice for more loads or containers than they have actually removed, nor to observe that these invoices are often paid without query because the client firm lacks any supervisory system and documentation to tell them the exact number of loads taken.

A variation on this scam will involve collusion between the contractor and the employee responsible for signing off each load. More loads will be authorised than have been taken, in fact, in return for a percentage of the value of each additional load.

Crimes involving collusion are difficult to identify, but one way of reducing this kind of risk is to ensure that jobs are rotated, so that any scam that develops from (in this case) regular contact between employee and contractor has only a finite term before the relationships are changed. An alternative is to require two records of any load moving off-site – perhaps one originating with the storeman in charge of waste and another separate docket raised by the security officer at the gatehouse. The two records would have to be reconciled and then matched with the contractor's invoice before payment is made.

STAFF PURCHASES

Many firms operate schemes whereby employees can buy the goods they produce at significantly discounted rates. This is recognised as being generally good for staff morale, offering, as it does, an employment *perk* to a whole range of employees who are ineligible for the company car or free private medical insurance.

However, such schemes need monitoring for potential abuse, and the wise employer may consider putting an upper limit on the amount any one employee can buy through such schemes. Employees in all kinds of businesses, ranging from toiletries to toys and from foodstuffs to furnishings, are likely to find a ready and regular market for products at prices cheaper than those available in the shops. Word quickly spreads of the availability of these 'bargains', so that what might have started as the occasional favour for family and friends can develop into a serious side-line and supplementary income for the unscrupulous employee.

Gerd Hoffmann tells a story to illustrate this that would be hilariously funny, were it not for the underlying serious nature of the events as they developed.

THE DAIRY WORKER AND HIS FRIENDS

It all started with the decision by a dairy products company to allow every employee one carton of milk free each day, to be drunk on the premises or taken home. One member of an employee's family preferred chocolate-flavoured milk, but because the ordinary milk was still required, the one carton a day rapidly became two. Soon cartons of yoghurt were added to the milk.

Then the employee took the decision to stock up on dairy produce prior to going away for the weekend to the site where he kept his caravan. Once there, neighbours were impressed with the employee's access to such perks and, keen to keep their own household expenses down, asked if he could help them out with cheap supplies as well. It emerged that one of the neighbours was in the butchery trade and, anxious both to impress and to facilitate the proposed arrangement, offered to provide by way of barter some choice cuts of meat the next time they met. So now the butchery worker is going to steal meat in order to barter for some stolen dairy products.

It did not end there, because an employee of the site supermarket learnt of the arrangements from the campers she served. She offered to provide some cut-price fruit and vegetables in return for dairy products from our original worker, and soon became thief number three as demand for her goods rose among the campers. Before too long, the supermarket manager investigated why his milk and dairy produce sales were down and discovered what was happening. He now wanted some of the cheap milk and yoghurt for his store, so that he could make a profit from the campers who were not within the bartering group.

The dairy worker was now having to resort to taking car boot fulls of dairy produce out of the factory every day in order to meet all these demands for his favours. He was able to do so mainly because, being low-value items that were produced in the factory, security crime prevention measures were virtually non-existent. Moreover, once the camping season was over, the co-operative circle decided that the bartering arrangements held too many benefits to allow them to lapse; so they agreed to continue meeting regularly through the winter months. As a result, what had started as the minimal exploitation of a staff perk was now established as an organised crime group involving three different thieves and a supermarket manager who connived with the arrangements.

Being in a position to supply goods at a discount on normal prices is often an ego boost to the employee thief. It enables him to claim that his position is more important than it really is, that it is his status that provides him with the opportunity to help others. The case illustrates how minor weaknesses and malpractices can develop into much more serious crime, from which it can be very difficult for the perpetrator to escape, even if he wishes to do so.

STEALING TIME

One way of wasting a company's money, without actually stealing goods, is not to do the work for which you are being paid. The most obvious candidates for this particular fraud are those who are only irregularly or loosely supervised, in which category the travelling salesman is pre-eminent.

A personal acquaintance in this line of work told me how, sometimes when he feels under pressure, or when he is at a loose end, he likes to drive his company car into the local cemetery, park it in the shade of the cypresses, put on a tape containing some soothing music, recline the car seat and peacefully reflect on the meaning of life for an hour or two. He says that he now gets eyed suspiciously by one particular gardener, who he feels regards him as a potential grave robber, or worse, but that the particular location offers one of the rare tranquil spots available to him in his largely city-based job.

Gerd Hoffmann tells of the sales representative who first did a little personal shopping, before driving to a local airfield, where he watched for a while the aircraft manoeuvres over a cup of coffee. For lunch, he travelled to a restaurant near the harbour, enjoying a leisurely meal in attractive surroundings, before moving on to a bar at the marina, where he had a drink whilst watching the sailing for an hour or so.

In both cases it is the company's time that is being stolen. Although it might be assumed that both people mentioned managed to satisfy their managers in terms of their overall performance, both clearly reveal by their actions how much further potential existed to improve on that performance.

PRIVATE SERVICE

The pattern of a service mechanic's work quite closely resembles that of the travelling salesman, although it is likely that he is expected to contact base more frequently. One of the most unpopular taxes for the consumer is VAT (currently 17½ per cent in the UK) that is added to the actual service charge. This gives rise to the not uncommon enquiry about whether there is any way in which the customer can avoid paying this tax.

Of course, it is impossible for the company employing the service engineer to avoid making the charge. However, the engineer might suggest that if help is needed on a future occasion, he would be willing to come along in his own time and undertake the work on his own account. In this case there would be no need for VAT, especially if payment could be made in cash. As an added inducement, the engineer might point out that he does not carry the same level of overheads that have to be charged by his employer, so the customer would gain all round.

From small beginnings like this, it will not be long before the service specialist is undertaking in his own time much of the work that would previously have gone to his employer. What is more, it is quite likely that the service work will be completed using the employer's van, tools and materials.

Both examples of time theft can only be resolved by management involvement and intervention, and it would be idle to pretend that there are easy solutions. The actions of both the salesmen and the service engineer represent a breach of the trust that should exist between employer and employee; but as a society we appear to be creating a workplace culture where mutual trust is disappearing, to be replaced by a series of short-term relationships dictated largely by the two parties' temporarily coincident self-interest. To a fair extent, both the salesmen's and the engineer's performance can be controlled and conditioned by performance targeting, although, as we have seen, this will not always realise full potential or prevent the erosion of the employer's customer base.

SALES VOUCHERS

A frequently used merchandising tool is the sales voucher offering some form of discount or bonus when used to purchase the specified goods or services. Less regularly used now than 20 years ago, but still available, are the gift stamps issued at the time of making a purchase and redeemable ultimately against goods offered in a catalogue. Current versions of the gift stamps are the tokens offered with petrol purchases.

Although the value of individual tokens is very small, there clearly is a realisable value, especially in large quantities. They do, therefore, need protecting from dishonest exploitation, of which the commonest examples are:

- Deliberate, clandestine overproduction at the printing works where they are produced, or the failure to destroy stocks that are slightly misprinted.
- The theft for their personal use by staff working in the factory or warehouse that services the catalogue operation, where tokens/vouchers sent in for redemption should be, but are not subsequently destroyed.
- Theft by staff working in the warehouse of the company promoting the vouchers, where stocks of vouchers are kept for onward distribution to the sales outlets issuing them to customers. The opportunity also exists for lorry drivers to steal sufficient quantities for their own use or to sell.

The controls that management can implement depend crucially on the recognition that these sheets of printed paper are, in fact, a form of cash – more akin to the Brazilian Cruzeiros than the German Mark, but cash none the less. Viewed in this way, they should be controlled in a similar way to banknotes and petty cash, counted regularly, and with usage balanced against receipts.

EXPENSES

Expenses have long been recognised as offering an easy opportunity for fraud, with the result that in most organisations they are usually checked. To an extent, management can determine the organisational culture in which fraud of this kind will either flourish or be generally perceived to be unacceptable.

An example of the lengths to which one substantial company went in its efforts to prevent inflated and fictitious expense claims can be appreciated from the fact that they employed two auditors permanently to scrutinise claims submitted by their own staff. These two people, for example, would telephone the hotel whose bill was the subject of a claim to check on the room rate, the cost of a *table d'hôte* dinner, and whether Mr(s) X (their employee) had stayed there on the date in question. They would check the best available fares with British Rail, the cost of a taxi ride from A to B, and the route mileage for car journeys with the RAC. The point is that every employee was informed that such checks would be made for claims they submitted, whilst the staff employment contracts specifically spelt out the penalty for submitting a wilfully false claim as dismissal.

Now, this was a large company, whose scale of operations made it possible to justify such thorough checks on purely financial grounds.

Despite this, one might question the overall merit of such an approach by pointing to the likely negative influence on staff motivation and loyalty.

Here are some general hints on things to look out for when scrutinising expense claims:

- *Hand-written receipts from petrol stations and restaurants, which usually issue printed receipts from the till.*
- *Test petrol obtained on the company's filling station account for reasonableness, compared to the miles recorded on the employee's monthly car log. Might his/her partner also be availing themselves of the facility?*
- *The same handwriting on receipts from apparently different sources.*
- *Vague descriptions of expense items, eg 'various', 'miscellaneous'.*
- *The absence of a precise date.*
- *Check carefully for the insertion of extra figures, eg converting £3.75 to £13.75. Do also remember to test for 'reasonableness', ie is this a reasonable amount to pay for this item?*

EXHIBITION SAMPLES

The concern here is over the difficulty of controlling the free samples that are usually provided to hand to those visiting the corporate exhibition stand. There is often a bartering trade among exhibitors' staff whereby samples intended for the visitors are exchanged according to what is available and which items take the staff's fancy. The problem can become severe if food and alcohol are provided for hospitality on such occasions, when bottles of whisky get exchanged for several six-packs of foreign lager or someone else's bottle of gin.

GOODS RECEIVED

This section will only look at one rather unusual scam among the many operated by transport drivers, because transport and distribution security has a chapter of its own (see Chapter 9). This particular operation is one uncovered during a Hoffmann investigation, but which will have a potential attraction for many and therefore represents a general vulnerability.

Buyers and Sellers Too?

During an investigation it was noticed that there was a lively trade at competitive prices in all kinds of items, usually promoted by drivers of independent distributors. Most goods were brand new and in their original packing. Further enquiries revealed that whilst some goods had been taken from the warehouse where they were held pending delivery, others were obtained by the drivers from employees of other companies to which they regularly delivered, and where they had established their unofficial trading circles. These items had been stolen by the employees to offer as exchange barter for the goods they themselves wished to purchase from the drivers. It was in the drivers' interest to expand their trade because the greater the choice of goods on offer, the greater the income they derived from the operation.

Is something similar occurring in your own organisation?

MANAGEMENT THEFT

Dishonesty of the kind described in this chapter is not confined to junior employees, but can be (and often is) practised by senior staff as well. Much the same can be said about frauds, where the relative lack of supervision to which managers are subject greatly helps them conduct the fraud. Whilst all this is widely recognised, there are some 'thefts' which result from managers' perceptions of their 'rights and privileges as a manager', and it is these with which this section is concerned.

Just Perks?

Growing stock discrepancies in a wholesale foods business, when investigated, were found to have their origin in the manager's interpretation of what was available to him as a privilege of his position. Every week he took his wife's shopping list down to the warehouse manager who supplied what was required from stock, laying it all out neatly in the boot of the manager's car. No attempt was made to hide what was going on, so it was not altogether surprising that other warehouse staff adopted the attitude that, *'What is good for the goose, is good for the gander'*, and fulfilled their own wives' shopping lists in a similar fashion.

A similar practice developed in a meat-processing plant, where the phrase, *'A quick sample for the laboratory'* had become a company byword. These were the words used by the manager every time he took a cut or two of prime meat from the conveyor belt, although it soon became known that the samples were never delivered for testing. The phrase was subsequently used jokingly by all the employees when they took meat for their personal consumption.

Both stories demonstrate the dangers of managers exploiting their position. They also point to the need to develop and publicise a company policy on staff purchases. In the first of these two cases, headquarters

*immediately instructed the warehouse manager to send to the account-
ant **all** lists of goods supplied to staff. He would price them at a conces-
sionary rate and then invoice the staff concerned.*

A BUSINESS WITHIN A BUSINESS

Although most employers would refuse to allow an employee to take a
second paid job of work, and many would also disallow employees to set
up and run a business of their own, it is not uncommon to find no specific reference to such prohibitions in employees' terms and conditions of
employment. Were this to be the case, it is likely to complicate any disci-
plinary actions an employer might institute when he wishes to end the
subsidiary employment.

Almost any category of worker might establish a second business, but
perhaps some are more likely to do so than others. We saw earlier, for
instance, how relatively easy it is for a service engineer to start up his own
list of clients. Car mechanics, painters and decorators – most skilled
trades people, in fact – would find it relatively easy to swell their earnings
in this way. Those with professional skills – lawyers, academics, accountants, and their like – might also exploit public demand for cheaper
access to scarce skills.

As the following case demonstrates, problems can soon arise from the
conflict of loyalties and the difficulty of maintaining a strict time demar-
cation between paid employment duties and those associated with the
private venture.

A wholesaler had employed Mr X as a representative for 30 years. He
had clearly performed satisfactorily, although managers noticed that his
recent level of sales had declined. Word reached them that Mr X
had been seen on his brother's business premises, following which
investigations revealed that for the last ten years he had spent four
days every week working for his brother, leaving only the fifth day to
telephone round all his employer's clients in order to generate new
business.

STOP AND SEARCH POLICY

Many of the cases given in this chapter will have illustrated the ease with
which company assets were removed from their premises. Many cases
crucially depended upon the perpetrators being able to leave the site in
the confidence that they were not going to be stopped and searched.
There is a great deal of technology to help make searching a surer and
fairer process (one only has to think of the checks made on airline

passengers), but access to the technology is futile unless the organisation has developed and established a policy for such an action.

It is clearly a sensitive issue and will almost certainly need the support of trade unions and staff associations if it is to be allowed to work effectively. However, we have seen some of the consequences of the absence of the right to stop and search employees, their bags and their cars, so it should not be ignored merely because it is difficult.

RECRUITMENT PROCEDURES

So far in this chapter we have examined some of the more common ways in which employees steal from their employers. However, the same reasoning that argues for the merits of installing good locks in stout doors, for CCTV surveillance and perimeter fences, because they are all designed to *prevent* crime, is for some reason frequently forgotten when it comes to making offers of employment. Yet the briefest pause for reflection is sufficient to realise how much better it would be to screen out potentially troublesome and dishonest job applicants *before* they are employed, thereby denying them the opportunity to steal from you.

Sensibly thorough pre-employment checks are therefore extremely important. The extent to which you check on an applicant's credentials and employment history will depend upon a number of factors, including the subjective judgement of how much you can afford to spend. However, a sensible approach to forming such a judgement is to assess the sensitivity of the job for which you are recruiting. Such an approach ignores the necessarily subjective evaluation of the person who has filled the post previously, and instead considers the potential harm to the organisation that would result from dishonest/disruptive/incompetent behaviour by the job occupant. Such an appraisal would consider issues like the financial, legal and ethical consequences of aberrant behaviour, take into account the efficacy of the constraints and controls surrounding the post and attempt to identify the degree of trust and authority that would necessarily be placed on the occupant.

At the conclusion of such a process, it might quantify the potential risk inherent in, say, the position of Managing Director of an autonomous overseas subsidiary company at £10 million, at which point the decision whether or not to spend £2000 on an independently conducted pre-employment check is a relatively easy one to take.

However, recruiting for such senior jobs is the exception rather than the generality; so what are the common-sense measures to take for posts of a more routine nature?

Written Applications

All job candidates should be required to complete a company-specific application form, and to answer all the questions it contains. Without a written record to rely upon, future arguments about what academic qualifications or previous employment experience were claimed will remain irresolvable.

Academic Qualifications

Job candidates realise how important academic qualifications are; indeed, the employer may well have specified those required for the post he is advertising. It is quite common for candidates to claim qualifications they have not obtained, whilst on the other side of the coin, there exist a number of spurious 'colleges' and 'academies' which sell qualifications, without the tiresome necessity of enrolling, teaching and assessing students first. If appropriate academic qualifications are an important dimension of the job, it becomes essential therefore that the employer sees the *original* certificates claimed by the candidate. Photocopies can be easily faked. With university qualifications, it is usually possible to get the awarding institution to confirm that the class and date of the award are as claimed. If doubts arise about the status of the institution whose award the applicant is claiming, calls to the appropriate local education authority, the Universities and Colleges Admissions System (UCAS), or the police station nearest to the given address are likely to provide an answer.

Previous Employment

If the job applicant *is* dishonest, more often than not their previous employment history will be falsified. The form which the falsification takes will vary, but will include claims for greater seniority and responsibility than was, in fact, the case; will extend starting and finishing times in previous jobs in order to disguise periods of unemployment (or imprisonment?), or to hide the fact that they have never stayed long in the same job; will explain occasional gaps in employment continuity by referring to periods spent working abroad; will offer periods of 'self-employment' for similar reasons; or will invent a period of employment with a company that apparently no longer exists, making it difficult for the new employer to check the facts.

References

Very few employers would dispute that it is vital to obtain references for potential employees *before* they are taken on. Yet experience shows that many organisations are slow to request them, are lax in ensuring that they do obtain them and are less than thorough in reading them. At least one of the references should be obtained from the candidate's most recent or current employer: a reluctance by the candidate for them to be approached might in itself be revealing. However, often it will be the case that a reference from the current employer may be understandably difficult to arrange before a job offer is made; if so, every effort should be made to obtain it speedily thereafter, and to make the job offer conditional upon a satisfactory reference.

Employers should be aware that dishonest candidates, finding it difficult to provide a reference from a previous employer, are not averse to giving a name and telephone number of an accomplice, as though this were the contact number for the Personnel Manager (or other office holder) at a previous place of work. The name given may even be that of the Personnel Manager. The candidate's accomplice will have been briefed about what is required, and will provide a glowing reference when telephoned by the new employer. To avoid being duped in this way, always check with Directory Enquiries that the number you have been given is that of the organisation claimed.

Police Records

It has been common practice to ask prospective employees for sensitive positions to obtain from the police a statement to show that they had no previous criminal record. This procedure operated under the terms of Section 21 of the 1984 Data Protection Act, through which any data subject had the right to ask a data operator for a copy of any data entry that related to them. The insistence by employers that potential employees obtain such data records from the police as a condition of employment became known as 'enforced subject access'.

The Home Office and the Data Protection Registrar are concerned to eliminate this practice, believing that it circumvents the purpose of the Act. Moreover, such data obtained from the police was seen to run counter to the Rehabilitation of Offenders Act 1974, in that it contained details of both spent and *un*spent convictions.

The Home Office proposed in 1997 that employers should instead apply to the police under the terms of the 1997 Police Act, which provides for expansion of the availability of police checks to include providing information on a job applicant's unspent criminal convictions. However, the procedures outlined seemed to be encumbered with excessive bureaucracy, giving rise to criticism during the consultation process. It remains to be seen how this question will eventually be resolved.

Psychometric Profiling

Several commercially available tests have been designed to reveal aspects of a potential employee's character, including honesty and predictors of future behaviour. It is fair to say that psychologists themselves hold different views about their effectiveness, with some extolling their benefits, whilst others argue about the risks of false interpretations or invalid results. For an overview of the subject, see Ira Somerson (1995); for a discussion of the legal issues raised by their use, see Peter Leeds (1994); and for a report of a successful application, see Gerald Borofsky (1993).

Use of Consultants

Readers should be aware that there are several specialist consultants who will screen job applicants. Most of us now appear on so many computer-held databases, that much of such screening can be efficiently undertaken by accessing these. The skill lies in knowing which of the several hundreds of databases to use. Those wishing to consider employing such consultants, but not knowing how to make contact or how to guard against charlatans, should ask a reputable professional association for a selection of names and addresses. The following organisations should be able to help. Their addresses and telephone numbers appear in Chapter 13:

- The American Society for Industrial Security.
- The Association of British Investigators.
- The Association of Security Consultants.
- The Institute of Professional Investigators.

References

Borofsky, Gerald 'Pre-employment screening for unreliable work behaviours: an opportunity to work co-operatively with Human Resource Managers', *Security Journal,* Vol 4, No 4, October 1993.
Hoffmann, Gerd (1988) *Tales of Hoffmann: The Experiences of an International Investigator,* Hoffmann Investigations, Amsterdam.
Hoffmann, Gerd *Hoffmann Detective Tips for Business & Industry* (bimonthly newsletter), Hoffmann Investigations, Amsterdam.

Hollinger, Richard C (1989) *Dishonesty in the Workplace: a manager's guide to preventing employee theft,* London House Press, Park Ridge, Illinois.

Leeds, Peter 'Legal concerns in the use of psychological screening tests', *Security Journal*, Vol 5, No 4, October 1994.

Somerson, Ira 'Pre-employment screening – an overview', *Security Journal*, Vol 6, No 2, May 1995.

Gerd Hoffmann's case material is derived from his book, *Tales of Hoffmann: The Experiences of an International Investigator* and from his bi-monthly newsletters, *Hoffmann Detective Tips for Business & Industry.*

Both are published by and obtainable from:

Hoffmann Investigations Ltd
Van Leijenberghlaan 199A
1082 GG Amsterdam
Netherlands
Tel: 00 3120 642 0237
Fax: 00 3120 642 5854

8

COMPUTER SECURITY

One business crime that has received a lot of media exposure is that involving computers. More specifically, 99 per cent of media interest has been concerned with the *black art* known as hacking, which is frequently glamorised and represented as the provenance of highly skilled youths exploiting the weaknesses in an advanced technology that the rest of us only imperfectly understand. In contrast to its sensational portrayal, however, this particular problem has not yet assumed an importance to justify this level of attention, whereas the much more common crimes committed within a computer environment have been underreported. One of the key purposes of this chapter, therefore, is to acquaint the reader with the range of criminal or otherwise improper activities associated with computers that are likely to be experienced within the workplace. Hacking will be mentioned, but there are many other more common problems.

The second purpose is to suggest a range of available countermeasures. As always with security, there is an up-front cost. What you choose to implement, therefore, should carry a cost that is appropriate to the level of the threat. There is no point in investing £1000 in security to eliminate a threat where the cost of an actual occurrence is only £600 – unless you expect to experience two or more occurrences within the life of the security measure. Having said that, however, it is vitally important that organisations dependent upon computer technology openly acknowledge that part of the cost of Information Technology (IT) is its security. Clearly, it is for *you* to make the cost/benefit assessment. What this chapter will do, if computer security is new to you, is to raise your awareness and indicate the kinds of issue to consider.

This chapter will first consider the range of threats to computer security, before revealing the kinds of countermeasures available. Within the threats section, there is a natural division between those of internal and those of external origin, whilst further divisions are possible according to whether the threat is the result of deliberate or accidental action. When we turn our attention to countermeasures, there are some that are applicable to both large and mid-range, centrally managed computers (mainframes) and Personal Computers (PCs), whilst others are more appropriate to one rather than the other. We shall try to keep these distinctions clear.

THREATS

A number of people have attempted to quantify the cost of breaches of computer security in the UK. Recent estimates have included one in 1994 by the Department of Trade and Industry of £1.2 billion during the two previous years, based upon its survey of 832 UK businesses. Another in 1995 by the PA Consulting Group (DeVille and Jenner, 1995) arrived at an estimated annual loss through computer-related fraud alone of £4–5 billion. At the end of the day, however, even the most ardent researcher is forced to admit that there is a large element of uncertainty in these figures because, it is commonly acknowledged, so much computer-linked crime is unrecorded. What can be indicated with some certainty is that the problem is widespread and growing in line with our increasing use of, and dependency on, computers.

One authoritative source of information has been the Audit Commission, which has produced six reports based upon surveys of computer fraud and abuse, the first undertaken in the five years to 1981, and others in the successive periods to 1985, 1987, 1990, 1994 and 1997 (Audit Commission). The Commission's surveys have aimed to assess the extent of fraud, theft, the use of illicit software, unauthorised private work, invasion of privacy, hacking, sabotage and virus attack occurring within corporate computing systems. Their data has been drawn from local authorities, NHS bodies, central government departments and agencies, a cross-section of other public-sector organisations and a range of UK businesses.

The report of the Audit Commission, *Opportunity Makes a Thief* (1994), drew attention to a threefold increase in the number of incidents in the three years since the previous survey, with 537 incidents reported in the 1073 participating organisations. More than one in three organisations had experienced an incident of one kind or another, which led the authors to conclude:

> *No sector is immune from computer misuse and the opportunity for fraud and other forms of abuse presents a very real threat. The total value of reported incidents has risen by a significant 183 per cent since the last survey, with an average financial loss for frauds per incident of £28,170. This, however, may not be a complete picture as the total loss from frauds is likely to be much higher. For some types of incident, the scale of direct loss will be more difficult to ascertain (invasion of privacy, for example) than for others (fraud and theft, for example).* (Audit Commission, 1994, p 9)

At this point, before moving on to a detailed examination of computer-related fraud methodologies, it is appropriate to offer a working definition. I propose to follow the definition used by the Audit Commission, that computer fraud is any fraudulent behaviour connected with computerisation by which someone intends to gain dishonest advantage. A broad definition along these lines allows one to consider the various manifestations of the problem without too much concern about whether or not the computer was essential to the fraud, or whether, by contrast, the computer was only incidental, in that the fraud methods could equally well have been used in a manual environment.

FRAUDS OF INTERNAL AND DELIBERATE ORIGIN

All the evidence from many surveys indicates conclusively that by far the greatest number of computer-related crimes are perpetrated by an organisation's own employees. The proportion revealed by the Audit Commission's research was 85:15 internally:externally originated events. Whilst a division of this kind has been evident for many years, the point has already been made (in Chapter 1) that recent changes in business organisation and ethos are likely to have increased the risk of inappropriate behaviour from middle-ranking, but professional managers. This threat is exacerbated, it was argued, when accompanied by the abandonment of traditional business controls in the search for operating economies. The Audit Commission makes much the same point:

> As organisations become leaner and fitter using technology to reduce layers of management, they run the risk of removing the controls and checks which former supervisory and managerial positions would have applied. **A key cause of fraud and abuse in this survey and in previous research was poor division of duties – an increasing trend to reduce control mechanisms for the sake of expediency.** (Audit Commission, 1994, p 12, author's emphasis).

So the concern is that there may be more internally inspired computer crimes in the future, not fewer, leading one to argue that this is an appropriate area to start an analysis of risk and exposure.

The most commonly recorded fraud methodology is the **unauthorised alteration of input data**, that is, changes made to the information being entered into the computer. These changes can involve amendments to genuine data, its suppression, or the creation of new, fictitious data.

Why should this be so common? Well, to carry out a crime depends upon a combination of technical knowledge and opportunity, plus, of course, the motivation. Within the specific sphere of computing there are many, many employees with a rudimentary knowledge of the technology – just enough, normally, to allow them to complete the particular tasks allotted to them. At the other extreme, there are only a few with the technical skills to allow them to write, for example, a virus program or a 'logic bomb'. Making changes to the information entered into the computer does not require a high level of technical competence, so it is easily achieved; whilst there are usually many people in the organisation whose job entails the routine processing of data. If the operation of the internal controls is slack (and we have suggested that there is evidence that this is not uncommon), then there is a situation in which many people with adequate technical skills will have been given the opportunity to commit a crime. All that is missing is the motivation.

The areas of business activity where this method is most often encountered are those where the consequence of the fraud can be converted into a financial benefit. This might sound obvious, but it is important to concentrate on how the fraudulent act can ultimately be converted into cash or goods. There are many examples from within accounts departments, where the perpetrators are already dealing with money transactions, but others occur within purchasing departments, stores and sales. Typical of the kinds of changes that are made are the entry of supernumeraries on to the payroll, the creation of fictitious invoices from suppliers (who might or might not actually exist), the erasure of the records of genuine debts due to the organisation, and the falsification of stock records to cover thefts of goods from the store.

Many of the frauds that have been discovered and recorded are, in fact, old-style frauds perpetrated within this rather new and imperfectly controlled computer environment. Most of them are not *high tech* in the way in which the newspapers commonly report them, but depend for their success on the exploitation of a weakness in the system by someone who knows and understands that system extremely well... and who knows the system better than those who operate it every day? Two examples of actual incidents will serve to illustrate some of the common methodologies.

This fraud was carried out by a 45-year-old accounts clerk, who had been employed for ten years. He withheld cash payments from customers that were received at the end of the month and did not post them as credits on the customers' computer ledger accounts. The customers concerned did not question the omissions because they assumed that the credits due to them would appear on the following month's statement, having merely been 'subject to delays on the computer system'. The accounts clerk did ultimately post the payments, managing to cover the shortfall caused by his theft of the cash amounts by withholding a further series of subsequent month-end receipts from that month's ledger statements. The ultimate loss was about £50,000 and was only discovered after two years because the clerk confessed to his superior. The motive appears to have been to support a gambling addiction.

The case illustrates two important points, both of which occur extremely frequently in the annals of computer-related fraud. The first is that the clerk faced a personal (and expensive) problem. In this case it was gambling; but at other times it is alcohol or drug dependency, the illness of a close relative, or an extra-marital affair. The second is the abandonment of that fundamental business control we have already mentioned – the need to keep separate the responsibility for recording the transaction of goods or monies from that of actually handling them. If the clerk had not been given responsibility both for receiving customer payments as well as recording them on the computer, the fraud would have been impossible without collusion.

One of the Purchasing Manager's tasks was to raise orders for the installation of equipment by outside contractors. He introduced fictitious items into the computer records of orders placed, which enabled the contractor to enhance the value of his invoices. These were sanctioned for payment because the system called for the invoices to be checked against the corresponding orders. The resulting overpayments were shared between the supplier and the Purchasing Manager.

Crimes involving collusion are difficult to detect, but in this case the segregation of duties control had been destroyed by allowing payments to be made after a reconciliation between order and invoice only, rather than on the approval of the Project Manager or Clerk of Works, following a physical inspection.

Entering fraudulent data into a computer system remains the most common means of defrauding an organisation, according to the Audit Commission research. When combined with the unauthorised amendment of computerised information, it represents 88 per cent of all computer fraud. As such, it is clearly a prime area for the attention of any manager wishing to reduce their risk exposure. Later in this chapter, we shall consider some of the available remedial measures, including the introduction and enforcement of the basic controls that so often seem to be absent.

Frauds that operate by amending computer data share many common features with those just reviewed, as the following cases illustrate. They commonly depend on the ability to circumvent access controls within the system or on the absence of effective controls. This allows the perpetrator to *browse* through computer files, a form of internal hacking, until an opportunity to defraud is identified.

A clerk whose job entailed recording and entering invoices for goods and services supplied by his employer contacted one of the major customers and offered to delete from the computer all relevant invoices, in return for a percentage of their value (DeVille and Jenner, 1995).

This event occurred in a multinational company which operated a computer network to link its various UK centres. The annual physical stock-taking by the local warehouse staff had revealed stock shortages of £500,000. This was serious enough, but what concerned management was that these losses did not show up as discrepancies on the computer-based stock-control system. The subsequent investigation revealed that a computer programmer from their Computer Services Department had been concealing these losses by writing down the stock levels shown on the computer records. He had done this by using a powerful utility *Edit* program to alter the stock balances on the central database, a use that was not recorded on the computer log. Two methods were believed to have been used: genuine stock movement transactions were altered, so as to inflate the quantities of stock apparently despatched from the warehouse and reduce the in-stock balances; and warehouse stock levels were arbitrarily reduced to cover the losses caused by thefts, whilst stock records for other company locations were increased by the same amounts, so that the overall UK control totals continued to balance.

Quite apart from the questionable wisdom of relying on local warehouse staff to complete the stock audit (with as long as a one-year interval), there is an important lesson here on computer security. As a good *rule-of-thumb,* never allow computer staff to have access to live corporate data. There is seldom a good reason why they should need this facility, but if a case has been made out, then make sure their activities are closely monitored and the details of their actions accurately recorded. Additionally, it is important to realise how powerful some of the utility programs are. The *Edit* facility is found on all substantial computers and allows the user to amend, delete or insert information on computer files. Often it is possible to use this same *Edit* facility to cover up actions such as the amendment of the stock records by deleting all references that would normally be recorded about them in the computer log.

Frauds involving computer output have traditionally been less frequent and less costly than those based upon the illicit amendment of input data. However, there are good reasons now to be concerned about this aspect of computer-linked crime, because of the portability of information held on computer floppy disks and the widespread decentralisation of computer printing. The major problem is likely to be one of keeping sensitive corporate information secure, rather than fraud, though an example of the latter is given below.

Information is one of the most valuable resources available to an organisation seeking to remain competitive within the global marketplace. Its usefulness is likely to be as great to a competitor as to the host organisation and it therefore needs protecting. It relates to finance, to business strategy, to research, to marketing and sales, to production processes and formulae, to purchasing terms and sometimes to government classified information (for example, as a defence contractor). Other information, whilst not so directly relating to the conduct of the business, either carries a legal requirement for security, such as that relating to individual persons (covered by the Data Protection Act), or would cause significant embarrassment if it was made public, such as details of salaries, share ownership or matters relating to the security of the organisation.

Because organisations are increasingly driven by computers, virtually all important information is processed and stored on them, from the Managing Director's monthly Board Report, to the Management Accountant's forecast of cash flows and capital and borrowing requirements, to the master file of customers and the discount terms they enjoy. One of the principal benefits of computers is that they can store huge amounts of information in a very compressed and economical form. In addition, the information can be copied and distributed extremely quickly via electronic mail, for example. It can readily be seen, though, that these advantages do at the same time represent a very real threat to security; because it is possible to store on one floppy disk, which will easily slip into a jacket pocket or a handbag, a copy of the MD's report, the Accountant's financial forecasts and the customers' master file. And if it is considered too risky to walk past the security officer on the gate carrying such a disk, then why not send the information by electronic mail direct to the computer on the desk of the competitor who has agreed to pay you handsomely for it?

So the recent concern has been how to keep information secure, as well as how to prevent frauds that depend on the misappropriation of computer output. Once again, we shall examine countermeasures later, but in the meanwhile, here are some case examples.

Virgin Airways claimed that details of its customer base and flight occupancies were hacked into by its major competitor, British Airways, part of whose computer was leased out to Virgin.

An international food retailer kept information on new products on PC floppy disks. Unfortunately, there were no access controls to the building, the office where the PCs were installed, or to the PCs themselves. An employee working in the R&D department of the same organisation made an illegal copy on to his own disk of the formulae for flavourings used in a popular food line and offered this to a former director, who was then working for a competitor. The former director took the disk to a computer bureau for printing, where an operator noticed that the name of the company that appeared on the disk was not the same as that which had ordered the print. She contacted the real owner and the theft was revealed (Wong and Farquhar, 1987, p 129).

A company involved in sensitive pay negotiations with a trade union was compromised when the union representatives produced some detailed information about their corporate finances that could only have been obtained surreptitiously. An investigation traced the leak to an employee working in the Computer Services Department, whose own home and part of his sister's were full of computer listings relating to the company's operations (Wong and Farquhar, 1987, p 130).

Temporary staff employed in a debtors' section used the on-line computer system to suppress recovery action over arrears on their own accounts and those of relatives and friends. They achieved this by suppressing the printing of the debt information, on which depended the subsequent action taken by other staff to recover the debts (Audit Commission, 1994, p 32).

OTHER INTERNALLY GENERATED CRIMES

These include frauds perpetrated by unauthorised insertion and manipulation of the program code, sabotage and vandalism, extortion, theft of equipment and theft of software.

It is understandable that crimes that depend upon the illegal use of a program code have gained a good deal of media attention. These are the *glamorous* crimes that require good technical skills to complete. In fact, apart from the particular manifestation of the computer virus, there have not been too many to come to light. Here is a short review of the genre, starting with the virus, which often originates outside the organisation, but usually depends upon internal activation to succeed. I feel that it is appropriate to consider the problem here, rather than in the section devoted to external threats.

The best definition of a **computer virus** is that coined by Dr Alan Solomon who said that a virus is a program that is capable of copying itself without the user intending it and without most users noticing it.

There are now hundreds of versions of viruses, ranging from the inconvenient and relatively harmless (the 'Italian', 'Stoned' and '1701' are early examples) to others which attack and destroy data files (of which the 'Friday 13th' or 'Jerusalem' virus is perhaps the best known). Some idea of the motivation behind many of the most harmful viruses may be gained by understanding that the name and first occurrence of this virus relates to the date of the 46th anniversary of the last day of Palestine's existence and the eve of the day when Israel declared its independence. The intention was to destroy all files held on affected PCs on 13 May 1988. Subsequently, every time the calendar produces a Friday 13th, there are fears that this particular virus will be activated. There is evidence of a growing worldwide industry of virus writers, especially in Eastern Europe and the Far East, from where the most damaging versions originate.

Viruses are now common: the Audit Commission reported that they accounted for 48 per cent of all the incidents recorded in its 1994 survey, with an average cost per incident of £977. To operate, a floppy disk containing a virus needs to be loaded into a PC where, through a variety of techniques, it is copied on to the hard disk. From here it will infect other software and data. Recent versions are very sophisticated and can lie dormant for a specified period or until a particular condition or event occurs. They may also be transmitted over networks if programs are copied without suitable controls.

Two recent and unusual examples of virus attacks serve to illustrate the potential for serious disruption that this quite common problem can demonstrate.

A worker at the Ignalina nuclear plant in Lithuania, at the time when that country was seeking independence from the USSR, attempted to introduce a virus into the plant's control system, but was foiled. The Tass news agency indicated that plant engineers were able to prevent any damage, but did not give details. Salius Kutas, the Lithuanian Minister for Energy, said that the subsequent shutdown of both reactors 'had nothing to do with the attempted attack', but was due to insignificant technical defects.

The owner of a design company in New York threatened to activate a virus in a customer's computer because he had only made a part-payment for the software supplied to him, after complaining that this was not up to specification. The threat claimed that the virus had already been installed by a technician and would be activated on a specified date if full payment were not received by then. The damage that would have resulted was estimated at $500,000. The potential victim was advised by the police to pay the outstanding amount, provided the virus was removed, but when a technician arrived to do this, he was arrested.

Discussion of viruses leads easily into the creation of other **illegal program codes**. Two of the best known are the *Logic Bomb* and the *Time Bomb*, whilst others can be classified as a version of the *Trojan Horse*, that is to say the creation of a secret facility within a program for subsequent exploitation by the perpetrator. The Jerusalem virus is a good example of a time bomb, in that it is triggered by the computer's internal clock when it reaches a predetermined future date. In simple language, the code says, 'When the time and date is X, take the following action...'. The logic bomb has a similar philosophy, but is triggered when certain predetermined events occur within the computer. Thus, a contract programmer discovered that when someone's name appeared in a certain file within the Personnel Department's sector of the computer, this was an indication that their contract was about to be cancelled. Because the programmer wanted to remain working with the particular client, he wrote some code that said, 'If my name (specified) appears in this file, then erase all the personnel data in this computer', specifying the files in which this data appeared. The 'Trojan Horse' quite often manifests itself as a device to circumvent the controls built into the computer program, typically to allow the perpetrator to commit a fraud.

As was indicated earlier, these are the kinds of crimes, along with *Trapdoors* and a specific technique known as *the salami fraud*, that the media loves to publicise. They do exist, but they should not be your main concern because they are rare. We shall refer again to trapdoors when discussing the problem of hacking.

We have already touched upon **extortion** in the New York case, but there have been others.

One similar event involved the insertion of some illegal code into the operating system, which caused frequent, if irregular breakdowns. The code is alleged to have been planted by a software engineer in a former client's machine, following the client's termination of his maintenance contract. The engineer indicated that the former client's computer was liable to fail and that the fault would be impossible to diagnose, without his help. The computer did fail and on the first two or three occasions the client rehired the engineer to correct the problem. However, he also put a team of skilled programmers to work to discover exactly how the breakdowns were triggered and they eventually discovered the illegal code, although only after significant costs had been incurred by the victim.

Another version of extortion has been witnessed several times and stems from the actions of disaffected staff. Some years ago, for instance, DSS computer staff went on prolonged strike in support of a pay claim, leading to severe problems over the preparation and distribution of pension and unemployment payments. This action appears to have had a profound effect on the government, who have subsequently changed their computing strategy from a central operation depending on a very large computer system to one that is substantially devolved to the regions. Here is a more recent example of the actions of disgruntled staff.

A computer operator and a programmer in separate incidents allegedly installed unauthorised software on their employer's computers to restrict access. One of the illicit programs encrypted database files and then decrypted them as they were called up on the screen. There was no visible indication to users that there was anything wrong, but, unknown to them, a time bomb had been installed that overwrote the decryption program at a specified date, after which all further files that were accessed from the database appeared on the screen in their encrypted form.

Vandalism and sabotage are other options for disaffected staff. Among recorded incidents is the man who took an axe to his employer's computer, an action attributed to his temporary derangement after he was bitten by an insect imported with some shoes from Brazil – a rather different kind of virus! There have been many cases of computer users deliberately corrupting information, whilst the following story clearly illustrates that *'There's nowt so queer as folks!'*

A recently installed mainframe computer broke down with monotonous regularity. The manufacturer for some time attended to these failures under the terms of their maintenance warranty. Eventually, however, they were adamant that it was not their hardware that was at fault. Suspicion ultimately settled on the computer night-shift operator, who was then subjected to covert surveillance. He was observed removing the rear panel of the central processor and deliberately causing a short that, of course, stopped the machine. Problem solved. But what was his motivation?

When questioned, he admitted to causing most of the other failures, but explained that he became lonely working nights on his own. He reasoned that if he caused a stoppage, he could legitimately call out the computer manufacturer's maintenance engineer, make him a cup of tea and have someone to talk to.

Finally, in our survey of internally generated crimes, we should be aware of the potential for the **theft of computer equipment and software**. One significant problem in the mid-90s has been the theft, not of the entire computer, but of one or more of the chips that drive it. This problem will occur again when we look at externally originated problems. But there was then a world shortage of chips; they had a high value and were easily portable. Ounce for ounce, I am assured that they were then more valuable than industrial diamonds. Furthermore, there are sufficient numbers of people with the skills to construct their own computer, or to enhance the performance of one they already possess, to indicate that this is a problem that is unlikely to disappear quickly. One indication of the theft of a chip is a reduction in the speed and performance of the computer from which it has been taken.

Theft of software is less of a problem now than it used to be, for more software is bought in than is developed in-house. Nevertheless, there is a great deal of pirated software in use, especially in the Far East. The Federation Against Software Theft (FAST) is actively seeking to protect the intellectual property vested in computer software and regularly brings cases to court where copyright has been breached. However, if your company has developed its own application programs in order to complete some task more efficiently, it is likely that it will have invested considerable amounts of money doing this. It is worth questioning whether this proprietorial software would have a value to competitors and, consequently, whether it needs securing.

> One major international airline was the first to develop an on-line passenger reservation system, having invested some 25 man years of expensive programming effort to this end. Just prior to going public, it is alleged that two members of this programming team took a copy of the final suite of programs and offered it for sale to a competitor airline. As a result, the market edge that this innovative work would have given the first airline was lost when its rival came to the market-place with the same product at the same time.

INTERNAL ACCIDENTS AND DISASTERS

One very common cause of computer-related problems – and a very significant cost – is accidents. Whether these are the result of poor quality controls resulting, for example, in the input of inaccurate information for subsequent processing; the malfunctioning of some item of equipment with only a tangential relation to the computer, like the failure of an air-conditioning unit; or a more serious and dramatic event, like a major fire, accidents account for a significant proportion of computer failures. A related problem is the occasional failure of computer software (whether bought in or developed in-company) to perform as intended or to specification. My favourite example is given below, but less dramatic occurrences are common, pointing to the importance of quality controls in the whole area of software specification and development.

> It is alleged that during the early trials of an American fighter aircraft, test pilots were faced with an unforeseen hazard. The plane was one of the first to pioneer the use of 'fly-by-wire' techniques, where computers were installed to help control the incredibly complex task of flying such a high-tech machine. Everything apparently went smoothly until the plane was flown across the equator, when it suddenly flipped on to its back.

THE MILLENNIUM BUG

There can be few people who are unaware of the computer software problem that has become known as 'The Millennium Bug'. Aware in general terms, that is – but clear about the specific nature of the threats to the world of business and services? Perhaps less so.

The Millennium Bug (or 'Y2K' as IT insiders prefer to call it) is actually a classic example of a logic or time bomb – albeit one that has been inadvertently programmed. It crucially depends on computers' ability to recognise the change of date from 1999 to the year 2000. In the days when computer memory capacity was limited, programmers saved space by using only the final two digits to indicate the year, rather than all four. This has been fine until now: but the prospect of a change of century seems bound to cause all kinds of confusion, with many computers unable to recognise what for humans is a simple and logical progression, and instead making the inference that *00* means *1900*.

The scale of the anticipated problems is staggering, because the phenomenon will extend far beyond mainframe and desk computers to all kinds of microchips in which are imbedded a date/time function. Thus, video recorders, ovens with timers, intruder alarms, process controllers, traffic lights, lifts, shipping navigation systems and many, many other everyday functions we take for granted are likely to be affected.

The seriousness with which the problem is now viewed can be gauged from the following selection of recent reports:

> *Oxford Economic Forecasting, the leading think-tank, has calculated that the costs of preparing computer systems for the millennium computer bug, and the disruptive effects of some computer networks crashing, will*

*slow world gross domestic product growth by 0.3 per cent a year. The fig-
ure sounds relatively small: but 0.3 per cent of the US economy is worth
$30 billion.*

*The study points out that infrastructure failures such as the breakdown of
electricity or telecom networks would be most costly. It also argues that disruption could trigger a stock market crash.* (*The Independent*, August 1998)

*'Developing countries' ability to tackle the year 2000 computer problem,
particularly in their nuclear industries, has emerged as a key concern
among officials of the Group of Eight leading industrial nations and the
European Commission.*

*(A UK government report) found that while large companies and organisa-
tions in the developed world appeared to have a good awareness of the situation, small and medium-sized businesses were less well prepared.*

*The problem was even more acute in the developing world. The report sin-
gled out the lack of preparations in India and Pakistan as being of particular concern, as were all the countries of Eastern Europe and south and
central America.* (*The Financial Times*, 6 August 1998)

*The Home Office confirmed yesterday that local authorities are being
encouraged to draw up contingency plans to deal with the 'nightmare scenario' of failed traffic lights, disabled water pumping stations, fuel shortages and other disrupted services.*

*In one key sanction, BT and Cable and Wireless have been told they will be
given the power to disconnect firms that corrupt phone connections (by failing to rectify Y2K faults within their own systems).* (*The Independent*, 11
September 1998)

*A third of pension funds have not received undertakings that their software
systems will cope with the date change after 2000, according to a report to
be published tomorrow. The report's findings come amid growing concern
about the ability of the financial services sector to prepare for the millennium bomb.* (*The Financial Times*, 10 September 1998)

*The insurance underwriters appear to have the clearest view of Y2K, and
their prognosis is bleak. Accordingly, insurance companies have already
decided to refuse cover for Y2K problems. The exclusion will affect most
types of insurance, but especially public liability. Airlines, public transport,
public buildings and retailers will be particularly vulnerable, since they cannot function without public risk cover.* (*The Independent*, 20 August 1998)

*But Y2K is more than just a problem of computer failure. It is about legal
dilemmas, insurance black spots, broken supply chains, public order and
contingency planning. The bug has exposed the fragility of a complex 'just
in time' economy.* (*The Independent*, 20 August 1998)

It seems certain that every organisation, however large or small, will be
affected by this problem in some way. Unfortunately, addressing and overcoming the problems tends to be time-consuming and (skilled) labour-

intensive, requiring as it does a detailed examination of the program code that drives the computers. In the UK there are government-led initiatives to stimulate action and to provide guidance: but at the end of the day it is down to individual organisations to provide their own solutions. A methodical approach will be based upon business impact analysis and contingency planning, which are topics considered later in this chapter.

THREATS OF EXTERNAL ORIGIN

The UK and other countries with advanced economies have suffered from the **theft of PCs**. In the early days, it was believed that these occurred because PCs were so desirable that a ready black market existed for them. To some extent that is likely still to be true, although the current *cachet* attaches more to lap-top computers, which are highly prized. However, PCs continue to be stolen, with those used in research areas, high-tech industries and universities featuring prominently. The pattern appears to indicate rather different motives from mere commercial gain from opportunist theft, and it has been suggested that there has been a concerted attempt by agents of Eastern European countries to gain access to some of the West's advanced technologies and research. This argument suggests that these thefts were inspired by the possibility of discovering useful technological information from the PCs' hard disks and, if it is true, reinforces the importance of keeping valuable corporate information secure.

We have already acknowledged the possibility of disaffected staff vandalising their employers' computers, but this and sabotage can equally well be the product of an external threat. Three recent **terrorist acts** – the two bomb explosions in the City of London and the bomb in the basement of the New York World Trade Center – all show the enormous disruption that can occur to business when buildings and the computers inside are taken out of action. In fact, the City of London is justifiably proud of how quickly and effectively it recovered from the second bomb; their spokesmen acknowledge how much they learnt from the mistakes they made during the first incident. In contrast, terrorism was a new experience for New Yorkers, and the long-term interruption to the businesses based in the World Trade Center was largely unforeseen.

> It's an ill wind, as Lord Hollick, the MAI boss will tell you. The bomb in the new World Trade Center in New York has hit the operations of Cantor Fitzgerald, the treasury broker based there. Cantor has been looking for temporary accommodation. Meanwhile MAI's Garban, one of Cantor's main rivals, is coining it. (*The Independent*, 16 March 1993)

Of course, not all disruptive acts are terrorist inspired. Great damage can be caused by **fire**, whether this results from **arson** or accident.

> Vandals broke a window and threw a petrol can and lighted match into the computer room during the early hours of Saturday morning. Two disk drives were burnt out, tapes had to be cleaned, the Central Processing Unit was scorched and data had to be reconstituted from back-up copies. Damage cost more than £150,000 and the smoke-damaged equipment had to be written off, as the manufacturers refused to maintain it after the fire. (*Computer Weekly*, 27 January 1983 and 16 March 1983)

Lightning strikes are not that uncommon in the UK, causing power surges that can severely disrupt computer functioning, and in particular networks.

> The Meteorological Office had just installed its new, very large mainframe computer in May 1988 when an electrical storm severely disrupted the service. Lightning strikes at the rate of one per second caused power surges that affected the disks, despite automatic cut-off devices and a back-up generator. The computer graphics were not available for the regular television forecasts.

Flooding and **earthquakes** have also caused computers to fail, recent examples occurring in Chichester and Kobe in Japan, respectively. With our increasing reliance on computer networks, any large-scale natural disaster is almost inevitably going to disrupt business computing. Having said that, there is every reason why we should endeavour not to make things worse for ourselves by failing to think through the consequences of our actions. Computers that are installed in basements, for example, or immediately underneath the water header tanks supplying the building, or (as in one truly infamous example) in a building under the flight path at the end of the main runway of a busy international airport, are clearly more susceptible to disruption than they need be.

However, the external threat that has captured public imagination more than any other is **computer hacking**. Hacking is the process of gaining unauthorised entry to a computer system, usually through the use of telecommunication facilities. Hackers are often computer enthusiasts with skills that enable them to move around networks and computer operating systems. The development of the Internet has increased the range of opportunities for them to learn of computer access codes, which are the prime means of gaining an entry into other people's com-

The Internet:Friend or Foe

When the Internet (formally the Arpanet) was implemented to connect various university campuses together in 1983, security was not a major design consideration. Its primary design was to allow anybody to read information and anybody to add information.

Today this produces some interesting security dilemmas. Much of our security has come from the statistical probability that we will not attract unwanted attention.

Information is the life force of all organisations and maintaining the integrity and availability of that information becomes ever more important. Today's systems must assure the integrity of their data and operating systems, they must have an expectation of privacy from their users especially when sending and receiving email and data files. They need to accomplish this across the Internet and other insecure networks (eg. telecommunications companies) to "untrusted" third parties.

Common risks associated with the Internet are:

- **Denial of service** (where key elements make the network or it's systems unavailable)
- **loss of confidentiality**
- **Viruses**
- **Unauthorised modification** (Trojan horse programmes)
- **Theft of service.**

The Internet has enabled user groups to freely exchange information to exploit corporation's and government's systems using all of the above as means of attack. However, the Internet is not the only area of threat: 80% of known attacks come from ex-employees or disgruntled workers.

For most companies controlling a user's access rights, data integrity and privacy both internally and externally can be difficult if there is no security management or security policy in place. The consequences of poor or non-existent security management are:

- **Corporate assets at risk**

- **High administration and monitoring costs**
- **Diminished user productivity from security access delays**
- **Security policies not enforceable enterprise wide**

This seriously diminishes the company's competitiveness.

How do corporations assess or analyse the danger of these risks and how do they produce and successfully implementing an effective enterprise-wide security policy over multiple systems and sites each with their own administration staff? The responsibility does not just rest with the IT Department; the security of a company's information ultimately lies with the Board of Directors.

How does the Board decide which levels of security are required across the enterprise to produce a cost effective security policy and more importantly how will they know if there has been a security breach?

Digital Pathways Services Ltd is one company that specialises only in Internet, Remote Access and Network security systems. It produces solutions that implement the company's security policy with cost effective, practical solutions that are user-friendly, centrally managed and independently audited. The following tools can all be used to distinguish friend from foe but more importantly it can keep your company in business securely:

- **Firewalls** can separate trusted networks from untrusted networks
- **Anti-Virus** products can eliminate known viruses from emails
- **Content blockers** will prevent confidential information from being transmitted to untrusted domains
- **Virtual Private Networks (VPN)** can protect user data and emails across public networks
- **Network Security Servers** will provide strong user authentication for remote users
- **Web blockers** will prevent user access to prohibited sites.

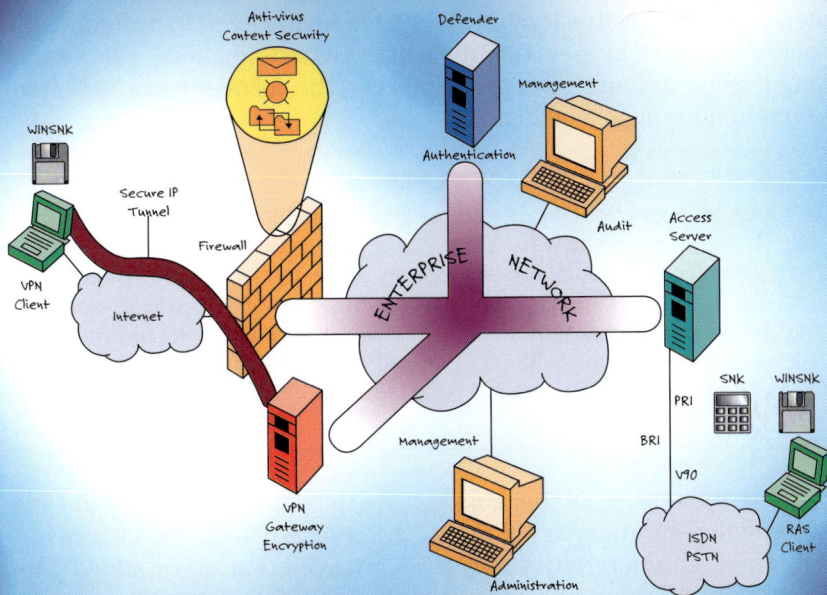

puters. Their motivation is often claimed to be the need to overcome the challenge posed by security measures, whilst financial gain is not thought to be a major incentive. However, the following example from the USA might suggest otherwise!

A computer hacker who blocked radio station telephone lines in the USA so that he could ensure he was always the winning caller to their competitions was jailed for four years. Keven Poulsen, 29, a former computer security consultant to the Pentagon, admitted fraudulently winning two Porsches, $20,000 in cash and at least two trips to Hawaii from stations in Los Angeles. He was ordered to repay $38,000. Poulsen is also facing charges of stealing a US Air Force list of war targets. (*ASIS*, 1995)

However, the real threat that hackers pose is to the security and integrity of proprietary information for, during the course of their illegal access to a corporate computer, they may have changed, erased or added data. The task of checking and reconstituting information that has possibly been compromised during a hacking attack can be extremely time-consuming and expensive.

The Audit Commission uses the National Health Service to make a very important philosophical point about what the future might hold (Audit Commission, 1994, pp 16–17).

Within the NHS there is a growing anxiety about computer centres which provide services to both medical schools and related hospitals. Medical schools want access to worldwide research and to encourage a free exchange of data, whilst hospitals want to protect patients' data and make it available only on a need-to-know basis. There is an obvious conflict where both use the same computing facilities.

This issue will become more acute as the NHS drives towards a national network connecting all purchasers and providers of health care. Whilst such technology will undoubtedly bring tangible benefits to patients, there are significant risks to data security which may prove tempting to the unscrupulous.

Many examples of hacking have received extensive publicity: the film *War Games* was based on it, whilst in the UK arguably the most important example was provided by the hack into the private mail-box of Prince Philip on British Telecom's Prestel service. This ultimately proved to be the spark that led to a change in English law with the advent of the Computer Misuse Act in 1990, which in essence makes it a criminal offence to obtain illegal access to a computer, whether or not any damage is caused to information or a further crime is committed as a result of the illegal access. The law is as applicable to employees who seek to reach computer information that lies outside their authorised access as it is to the external hacker.

Readers interested in gaining a more detailed insight into hacking techniques are recommended to obtain a copy of *The Cuckoo's Egg* by Clifford Stoll (1989), which is both informative and an extremely enjoyable read, in the genre of a thriller.

COUNTERMEASURES

Business Impact Analysis

It is important to understand the nature and extent of the threats to computers in order to focus with purpose on ways of countering the problems. Moreover, an essential prelude to the introduction of any countermeasures is an analysis and evaluation of the risks the manager faces in his particular environment. So far, we have indicated a wide range of both criminal activities and accidental causes of computer losses, but it is unlikely that any two corporate profiles will be identical; the result is that the emphasis in any defence strategy has to be a matter of individual calculation. The suggestions that follow need to be viewed in this context.

Risk analysis was considered in Chapter 2, where it was seen to be an important tool in any strategy to control losses. It is just as applicable to computers and information security as to the physical security of premises. There are some risk-assessment tools to help the analysis process (Computer Risk Assessment Methodology (CRAM) and Dependency Modelling are two), but the key to giving computer security an appropriate focus is to undertake a **business impact analysis**. By this is meant the process of asking a series of *'What if...?'* questions, based upon three main considerations. These are:

- What is the potential impact on the organisation's day-to-day business and financial capabilities, if a security breach occurred in this particular area?
- What are the financial costs likely to be if a security breach occurred of this specific kind?
- What would be the value of the loss of goodwill and business credibility that might result if this specified breach in security occurred?

The associated question the manager must then ask is:

- How great is the risk that this event (which I have identified as likely to have a significant impact on the business) will actually occur?

It is by answering these questions that the manager can prioritise computer applications for security *in terms of their importance to the business*, whilst, by assessing the likely costs of *in*security, he has also identified a yardstick for regulating his investment in whatever security measures are identified as available to reduce the risk. Contingency planning of this kind is no mere academic exercise, nor should the associated costs be viewed as another example of an unnecessary corporate overhead. The earlier part of this chapter demonstrated how wide is the range of risks to computer operations and, given the level of dependency on IT that is now universally achieved, security should be regarded as a key management strategy.

Currently, there is no more important application of this approach to managing IT security risks than for approaching the problem of the Millennium Bug. It is the philosophy behind Action 2000's™ approach to tackling this time-critical problem. Action 2000™ provide a free action pack to help small and medium-sized enterprises analyse their exposure to the risk of disruption from the Y2K bug, and the nature of their approach can be gleaned from the questions they pose in the publicity material:

- Have you called Action 2000 on 0845 601 2000 and asked for your free Action Pack?
- Have you informed your staff about the issue and identified who will be responsible for getting your business ready?
- Have you drawn up a list of all the computer-based systems, components and electronic equipment that support your business?
- Have you ranked these items in order of importance, based on which ones will cause the most disruption if they fail?
- Have you decided whether you will repair, replace, retire or upgrade items that are not Millennium-ready?
- Have you undertaken a comprehensive 'live' testing programme to make sure the systems you've prepared are reliable?
- Have you ensured that all your customers and suppliers have taken similar steps before relying on them?
- Have you developed contingency plans in case of unforeseen problems?

This is a good, practical example of how business impact analysis can be used both to identify problems and then to prioritise them for attention and resolution. Moreover, within the broad framework of risk assessment there are many security options. Some of the more important will be considered next.

Policy and Code of Practice

The first security option is more an essential prerequisite than an option. It concerns the need to develop and publicise a **computer security policy**, which need not be long or detailed, but which will:

  indicate the extent of management's commitment to security;
  set objectives;
  define the responsibilities of staff;
  refer to any legal or contractual requirements that impinge upon security;
  indicate how the policy will be monitored and controlled;
  indicate the consequences for staff who disregard the policy.

This last item will not necessarily prevent people from misbehaving within the corporate computer environment, but if it forms part of the employee's terms and conditions of employment, it is likely to avert the unpleasantness and inconvenience of any industrial tribunal that an employee might otherwise contemplate invoking.

Following on closely from the statement of policy is the development of a **code of practice** for computer security, which should define standards of security in sufficient detail to allow business managers to develop procedures to translate policy into action within their own areas of responsibility. Fortunately for us in the UK, all the hard work involved in developing a suitable code of practice has already been completed by a group of managers from some of the country's leading businesses, with the support and patronage of the Department of Trade and Industry. They worked to the following objectives:

1. To provide a common basis for companies to develop, implement and measure effective security management practice.
2. To provide confidence in intercompany trading.

The results of their labours is published as *A Code of Practice for Information Security Management* and has now been adopted as British Standard 7799 (1993). This provides a comprehensive, accessible, business-oriented and jargon-free guide to good practice and should form an essential starting point for anyone concerned about IT security. Its broad headings are:

  Security policy.
  Security organisation.
  Assets classification and control.
  Personnel security.
  Physical and environmental security.
  Computer and network management.

System access control.
System development and maintenance.
Business continuity planning.
Compliance.

Fraud Prevention or Reduction

Whilst one's main aim should be to prevent fraud occurring, it is unlikely that anyone who wishes to be able to continue trading could create a wholly risk-free environment – the overheads in terms of operating inefficiencies would simply be too great. There is therefore also a need to give early warning of fraud, to limit the exposure and to provide the means of detecting what happened.

A form of control that is often only weakly applied in businesses is that of **good personnel procedure**, by which is meant the application of thorough pre-employment screening for permanent staff. Nowadays, the increasing reliance on contractors, whether short or long term, poses an additional problem for employers. Who undertakes the checks on such workers – the agency or the employer? And where will the contractor work next – for a competitor?

One prime business control is the **separation of duties**, making which is meant eliminating the possibility of any one individual completing a business transaction from beginning to end. Within an accounting environment, this translates as segregating the task of recording a transaction from that of handling the actual goods or monies. For such a fundamental control, it is truly amazing how often frauds occur that depend wholly or in part upon its absence. Somehow, in the process of moving work on to a computer, managers seem to lose sight of its importance. Recent pressures to delayer organisations and to search constantly for operating economies might also have contributed to the problem, by eliminating posts and reducing in particular the numbers in supervisory roles.

We have seen that most frauds are perpetrated by employees making illicit amendments to information being entered into the computer or by tampering with the output. It is therefore very important to control these stages in computer processing. **Controlling input** involves establishing both physical and logical access controls to terminals and work stations, enforcing in particular the secure use and regular changing of passwords, which remain the most common method of logical access control. Easy-to-guess passwords should be prohibited, and it is possible to program a range of the most common into the computer so that potential users of them are debarred from the start.

Nowadays, passwords can be backed up by additional controls, like electronic one-time password generators. Sensitive information can also be protected by a second level of password and can also be stored in encrypted form, making it impossible for an unauthorised viewer to understand unless they also discover the decryption key. In order to be able to unravel a suspected input fraud after the event, one needs a tamper-proof audit trail. Precautions such as these will be largely circumvented, however, if inadequate physical access controls to the building and the computer enable an unauthorised user to walk up to a machine that has been left running whilst the logged-in employee is taking a tea-break, for example.

Job rotation, where possible, is an effective way of reducing the opportunity for employee fraud, since so often this depends upon the exploitation of a perceived weakness in the procedure which is being operated. If the work environment is changed regularly, not only will you reduce the likelihood of any weakness being identified, but you are also likely to enhance work satisfaction within routine tasks by achieving variety.

Paper authorisation procedures, such as those which then serve to generate weekly allowances more or less automatically, are a vital control for any amendments to master files. They are equally important to control program and system changes made by IT staff.

Inhibiting **output fraud** is possibly best achieved by implementing measures designed to protect sensitive corporate information, since it is the loss of this that represents the major threat today (see also Chapter 6). Among the many possible actions, those with particular relevance to computer-linked fraud include close control over printers and the general distribution of printed output, as well as sequentially numbered cheques, and confronting the difficult issue of taking corporate information off the premises so that the employee can work at home on his portable computer.

The timely **detection of fraud** depends largely on close monitoring and surveillance for compliance with the established procedures and systems. Clearly, it will help if there are internal auditors to undertake this function, but a good idea of potentially fraudulent behaviour within the computer environment can be obtained from a well-conceived and monitored **computer log**. More correctly termed the 'activity transaction monitor', this is a record of what has happened on the computer – who has logged in, at what time, from what terminal, to what file, and so on. Quite clearly, most of any day's transactions will be entirely legitimate and, from a security point of view, therefore, totally uninteresting. The log itself, on any central service computer, will be an extremely lengthy document, *unless* it is restricted to recording only those events which have been identified previously as unusual. You can program the log on an 'exception reporting' basis so that only unusual events, identified previously as likely to have an implication for security, are recorded. You might choose, for example, to 'flag' access to any particularly sensitive files; to record any log-ins that take place out of normal working hours; to

Telecommunications Security

The following was written by Praesidium Services Limited, an international security consultancy specialising in telecommunications fraud management.

Tel: +44 1249 467 800
Website:
www.praesidium.com

Fraud on telecommunications networks is a worldwide problem affecting all providers of telecommunications services. It is estimated that the global cost of telecoms fraud is well in excess of £10 billion per annum. What is less well known is that all companies which have a PBX, a private branch exchange otherwise referred to as an internal telephone system, could also be exposed to the same fraud problem. Telecoms fraud can be defined as the obtaining of telecoms services by means of dishonesty with no intention of paying. Telecoms fraud exists for many reasons, because the ability to speak to people in other countries for free or very cheaply has become a desirable international commodity.

Why does telecoms fraud exist?

There are a number of factors which determine the scale of telecoms fraud in a particular city or country. The existence of call selling operations all around the world is without doubt the single biggest contributor to the overall cost of telecoms fraud. A call selling operation is a fraudulently organised criminal activity, whereby unauthorised access is gained to telephone lines, which is then sold on to other parties. The demand for these type of fraudulent services is widespread, as most people have family or friends living in other parts of the world and many could be attracted by the possibility of making long calls at rates substantially below the incumbent telecom operators' international calling rates.

Some telecom fraudsters don't attempt to make a business out of setting up a call selling operation, but will merely exploit the opportunity for their own purposes, in order to save themselves money on international calls. Furthermore, many other types of international criminals are not only attracted by the ability to make free calls, but also by the anonymity that telecoms fraud affords them. By using someone else's telephone line, a criminal could evade detection from the police authorities in a particular country.

Another major fraud risk which can affect any business is the exploitation of premium rate service (PRS) fraud, which occurs when excessive calls are made to expensive PRS numbers, which are owned by the fraudsters. The Premium Rate Service Providers rent these lines from one of the incumbent operators and then would fraudulently inflate the inbound traffic to their own services by organising other parties to call one of their numbers. The risk is that anyone else's line could be fraudulently utilised to make these types of calls. This would allow the fraudsters to significantly inflate their revenues due to be paid to themselves by the telecom operator.

Lastly, there is also a risk of telephone hackers committing fraud not for any financial gain, but merely to be able to prove that they have the ability to access other parties' telephone lines. The main threat with this type of activity is that hackers enjoy sharing this type of information over the internet, which is then widely available for others to utilise.

How is telecoms fraud perpetrated?

Fraudsters have many possible options when deciding how to commit a telecoms fraud. Essentially they will either fraudulently subscribe to a service with no intention of paying or they will technically gain unauthorised access to another party's line with the same intention of not paying. The former is in fact the oldest and most common type of telecoms fraud prevalent around the world and is called subscription fraud. This type of fraud affects all types of telecom operators, whether they are selling fixed-line services, mobile phones or even calling cards.

However, the latter type of technical fraud is the one that represents the key risk to all companies operating a PBX. In addition to being able to compromise a telecom operator's networks and systems, the fraudster will also attack any unsuspecting company's PBX. All companies' PBXs are connected to the public switched telephone network in that country, which in turn is connected to the

international networks of those telecom operators offering international connections to other countries around the world. Therefore, by gaining unauthorised access to a company's private PBX, the telecoms fraudster could set up international calls around the world at the company's expense.

How can PBXs be compromised?

PBXs usually have a facility whereby an outside caller can access the system for convenience of use, maintenance or management purposes. This type of feature is referred to as Direct Inward System Access (DISA) and the process of permitting an incoming call to connect onwards to an outside line is called Breakout. The reason that DISA is widely available as a PBX feature is because many companies encourage their staff to utilise the feature in order to reduce the corporate telephone bill. For example, if an employee is staying overnight at a hotel and wishes to make an international call, instead of using the expensive hotel phone, a call can be placed through the company's PBX. Similarly, this feature could also be used when an employee is working from home or if he is on his mobile phone, which may have international dialling access denied.

The DISA service can be used by dialling a dedicated PBX extension number and inputting a personal authorisation code, followed by the required telephone number. The company's PBX would then dial the inputted number in order to set up the call. In addition, some companies' mobile phone users may be able to utilise this type of feature without the need for inputting an authorisation code or PIN. The expansive use of voicemail systems operated off the company PBX also creates a fraudulent opportunity. These systems are generally designed to allow employees to call in from anywhere to access their voicemail. Upon dialling their extension number and inputting a PIN, the messages would be played. After hearing the messages, many voicemail systems would allow an outside call to be placed.

Despite the security protection of authentication codes and PINs, the PBX breakout and voicemail features can sometimes be accessed by fraudsters. The most frequent method of attack is for the fraudster to dial into a PBX and to test the most common default codes or passwords. Intelligence can also be gained about the different default code settings for all the PBX manufacturers' models. Some fraudsters will utilise sophisticated automatic diallers to test a huge amount of possible codes in an effort to identify a valid one. Furthermore, another option for the fraudster is to attack the PBX's maintenance port, which would enable the altering of the system configuration. This type of attack is probably the most dangerous for a company because the fraudster could establish himself up as a new user, set up a permanent call forward to another number or even suppress all call records for a specific extension. Lastly, there is the possibility of "social engineering", where the fraudster calls up someone in a company, possibly a switchboard operator, in order to ask them how specific features can be set up or altered. The fraudster will typically be very skilful at extracting information over the phone.

Identifying the vulnerability of a company's PBX

The first issue to resolve when considering the risks associated with a particular PBX is to establish whether a DISA facility is present, is it being utilised, or has it been disabled? Similarly, can follow-on calls be set up from the voicemail system?

The next critical issue to establish is does the PBX have a remote access port for maintenance purposes? How is this facility controlled? What activities can be undertaken over this link? And have any security mechanisms been put in place? Are regular reports produced which can monitor all PBX traffic?

Most importantly, is there a dedicated person responsible for the company's telecoms requirements? And does this individual include the security of the company's PBX and other telecoms services as part of his responsibilities?

Guidelines for securing a company's PBX

All companies should consider introducing a range of fraud prevention and detection measures in order to reduce their exposure to telecoms fraud perpetrated through their PBX. The following details the key issues which should be included in the company's PBX policy and procedures manual:

All information about system access procedures as well as internal telephone directories should be treated as confidential, not only in terms of storage but also during disposal. Staff should be made aware of the confidential nature of this type of information, especially the switchboard operators.

The default access codes should be changed as soon as possible and common codes should also not be used across the company. There should be written procedures about permanent and temporary staff leaving the company, in order to ensure that their access codes are immediately changed or deleted and that their mailboxes are locked.

DISA access should not be available to all staff, but only a "need to have" basis and a register of users should be kept and updated. The number of staff given access to the maintenance facility should be restricted to very few and the remote access ports should be constantly disabled unless the system requires servicing.

Access codes should be changed frequently, ideally monthly, and staff should be encouraged to use the maximum number of digits allowed by the system. Password management features, which force users to change their passwords after a specific period of time or after a number of access attempts, should be deployed.

Company policy should prohibit the use of the telephone number as the access code. DISA should never be possible by dialling any corporate freephone number or information line.

The system's activity reports should be regularly monitored, ideally daily but at least weekly, in order to identify any unusual calling patterns for any particular extension. The system should be configured to restrict the number of invalid access attempts and to create an alert if there are attempts at unauthorised access.

Consider utilising call restriction features which would facilitate the barring of DISA traffic to nominated number ranges, such as premium rate numbers, specific countries with a high fraud risk, or possibly whole continents. In addition, special features such as call forwarding could be totally removed.

Telephone Card Fraud

In addition to the risks of fraud over a company's PBX, the risks associated with charge cards and calling cards should also be recognised. These cards allow staff to make calls when out of the office or even abroad; charge card usage would be included in the total bill charged to a specific telephone number, whereas calling cards would be settled directly either by a monthly statement or paid in advance. The access codes to these cards should be secured in the same way as the DISA codes. However, there is an additional risk of card counterfeiting, where the fraudster finds out these codes and then uses the card's identity to make unauthorised calls. Tactics adopted by card fraudsters includes observing people using their cards to make a call and then memorising the access codes and, in some cases, stealing call loggers from hotels which register these codes when the cards are used by hotel guests.

Mobile Phone Fraud

The mobile phone is often viewed as the ideal solution for staff when on the move. The most common type of mobile phone in circulation is GSM, which requires the user to have a GSM handset as well as a SIM, which is a smart card. GSM is a secure technology and as such is not exposed to cloning problems in the same way as the older analogue networks. There is very little chance of any calls on an employee's bill not having been made by that person or someone else who had access to his mobile. Therefore, the only fraud risks to which a company would be exposed, is the possible theft of its employees' GSM mobile and its subsequent misuse to make unauthorised calls, together with the cost of replacing the stolen handset. Staff should be made aware of their responsibilities to look after their mobile phone and also to immediately report any incidence of theft.

Employee Abuse

The distinction between fraud and staff abusing their telephone at work, their DISA access, telephone card or mobile phone can often be unclear. Access to "free" telephone service is a significant temptation, which can sometimes lead to abuse. As such, there should be clear guidelines as to what type of usage is considered acceptable and these should be communicated to all staff. Similarly,

regular monitoring of the PBX traffic for fraud purposes would also reveal any potential staff abuse.

Staff abuse could also lead to fraud if an employee decides to exploit his knowledge about the company's telephone system and then advises other parties how to access the system. The personnel department should be consulted when developing a company policy on the use of telephone services.

Concluding Remarks

In spite of telecommunications being a significant cost for all businesses today, many companies have not nominated a dedicated person to be responsible for all the business' telecommunications requirements. Furthermore, in most circumstances where such a person is appointed, insufficient attention is applied to the inclusion of security as an integral part of his role and responsibilities.

The company telecoms manager should develop a telecoms security policy, together with supporting procedures, which should address all the issues covered in this article. Escalation procedures should also be included in order to respond to the different types of security incidents which could occur. Most importantly, the telecoms manager should constantly monitor all telecommunications traffic in order to identify any fraudulent incidents or new trends. In this way, a range of prevention and detection measures could be developed across the company, which would enable financial losses due to telecoms fraud to be minimised.

note any multiple attempts to log on; and to match users with their customary terminals and to record any activity that takes place outside this norm. This way of using the power of the computer as a positive aid to security has much to commend it, but to be effective, someone must be assigned responsibility for at least daily scrutiny of the log *and for initiating any follow-up action.*

Viruses

The golden rule is to have a copy of everything that is important. Some viruses attack and destroy data or deny access to it, so it is vital to have a back-up. Although it will not save you in every circumstance, it should do so in 90 per cent of cases.

Secondly, get all your software from a reputable source – bulletin boards certainly do not fall into that category. Another example of a disreputable source is your acquaintance from the firm next door.

The third rule is to write protect all disks to which you do not need to write. (You can achieve this by taping over the moveable cover on the floppy disk, so that the write head in the disk drive cannot access the floppy disk surface.) The most common viruses then simply cannot be acquired, because they cannot attach themselves to your programs.

Fourthly, you should buy some anti-virus software, some of which is *virus specific* (that is, it looks for known viruses) and some of which is *virus non-specific* (in that it searches for any changes in the format of your proven software).

Finally, you should ensure that you include a statement about viruses in your computer security policy and then promote a good awareness of viruses among all computer users. New software should be tested on computers that are isolated from any corporate networks before adopting it for general use. Taken together, these measures should prevent you from becoming a virus victim.

Hacking

Although hacking incidents are, mercifully, still rare in this country, it is worth bringing your attention to one or two safeguards against the threat. First, if you discount any activities undertaken by staff from within the organisation, hackers need to access your computer by a telecommunications link. A computer that literally has no such links with the outside world, therefore, is not vulnerable to hacking. However, most do, and therefore the first thing to address is the security of these links. If hacking is considered to be a serious threat, then it is probably worth investing in a 'sentry', an electronic device that intercepts all incoming calls to the computer and asks for the user's ID and password. It then disconnects the call, refers to its own security tables to check if the ID and password are valid, and then dials back to the authorised number associated with these. Clearly, although a hacker might have guessed a legitimate user's ID and password, it is extremely unlikely that s/he will also have access to that person's authorised telephonic address.

Apart from this, it is good to be aware of the practice of computer manufacturers to provide a dial-up line for use in fault diagnosis and maintenance. Such provision allows their engineers to attend to any faults on customers' machines from their own work space, rather than having to travel to clients' premises. They claim that this is cost-efficient, but it is also a trapdoor through which any hacker can gain access. If the host computer is also linked to others within a network, the possibility exists for the hacker to move around within this enlarged environment.

A similar trapdoor sometimes results from the computer supplier's installation procedures. Many installation manuals contain a number of privileged user identification codes and passwords to the system. In the film *War Games*, the computer experts that the young hacker called on for advice referred to the 'back-door' access facilities that they had planted in the computer systems they had commissioned. Unless the customer's own installation team make a conscious effort to disable such facilities, a knowledgeable programmer, a hacker or even a user could employ these default codes to gain access to corporate data by circumventing the security controls.

Finally, do enforce the use of passwords that are difficult to guess and do not give users the opportunity to employ simple, short passwords. A minimum of eight characters, an avoidance of common words and the insistence on a mix of alpha and numeric characters will make it that much more difficult for a hacker to crack. Of course, all these good intentions will be fatally undermined if the hacker gains access to the central file of passwords and can read them there for all users. To prevent this, make sure that passwords are one-way encrypted before they are stored.

Contingency Planning for Disaster Recovery

We identified earlier a range of events that might cause a prolonged disruption of the computer service. Because of our extreme dependency on computers for the conduct of our businesses and services, it is barely possible to contemplate a form of working where we were denied access to computers for any length of time. Reversion to manual methods is simply no longer a viable option.

This dependency in turn means that we must address the issue of how to maintain business continuity in the rare, but still eminently possible, event of becoming the victim of a major incident that results in our computers being unavailable for an extended period. Contingency planning should start with a business impact analysis, a technique already mentioned in this chapter. This will enable the business to prioritise computer applications for recovery and it will lead to a formal consideration of how that recovery might best be achieved.

It is sufficient at this stage to mention that various recovery options are available. The most expensive involves the installation of a second computer running in parallel (but perhaps at a remote site) with the first, and able to take over the processing immediately this is put out of action. Other options involve what is essentially an insurance policy that provides access variously to a fully equipped computer service, able to take over your processing at minimal notice; a mobile service delivered to your own site; or empty office space, available to you for installing a replacement computer. Known colloquially as 'hot', 'warm' or 'cold' start facilities, such services are now commonly available. Because the facilities are shared with other potential users, they offer a cost-effective insurance against major incidents.

With contingency planning, there has been a tendency to overlook the need to make adequate provision for the work processed on PCs, whether these are in stand-alone or networked form. The need to reconstitute the telecommunications links themselves is also liable to be overlooked. (A fuller examination of these extremely important issues appeared in four consecutive issues of *Computer Audit Update*, 1995.)

CONCLUSION

In this chapter we have given computer security a good airing. This is justified because of the extent to which computers now provide an essential tool for management in the efficient conduct of their businesses. In the past, there has been a danger that security within this still relatively new technology has been ignored, through a combination of an imperfect understanding of the risks and solutions on the part of the business managers, and a completely different set of priorities among the technically proficient IT managers. These latter may well have had access to the technology to provide a technical 'fix', but have traditionally been much more concerned about getting new applications commissioned to a timetable and, in addition, may not have been the most appropriate people to prioritise business risks. The major purposes of this chapter, therefore, have been to eliminate any remaining mystery about the technology, to raise awareness of the risks and criminal methodologies, and to place the issue of computer security firmly within in the area of business management.

References

ASIS, the newsletter of ASIS (American Society for Industrial Security) UK, Chapter 208, spring 1995.

Audit Commission (1994) *Opportunity Makes a Thief,* HMSO, London.

For more information on the computer fraud surveys and the work of the Commission, please contact The Audit Commission, 1 Vincent Square, London SW1P 2PN. Telephone: 0171 828 1212.

British Standard 7799 (1993) *A Code of Practice for Information Security Management,* DTI, London, and BST 1993.

Computer Audit Update 'Business Continuity Planning', by Keith Hearnden, Elsevier Advanced Technology, Oxford, May, June, July and September 1995.

DeVille, Tim and Jenner, Peter (1995) *The Role of Information Technology in Combating Fraud,* PA Consulting Group, London.

Stoll, Clifford (1989) *The Cuckoo's Egg: Tracking a Spy Through a Maze of Computer Espionage*, The Bodley Head, London.

Wong, K K and Farquhar, W F (1987) *Computer Crime Casebook*, BIS Applied Systems, London.

9

TRANSPORT AND DISTRIBUTION SECURITY

PROTECTION OF ASSETS

Finished product, raw materials, work in progress, plant and equipment, and the vehicles that transport them are considerable and precious assets to any organisation. They are never more vulnerable to theft than when they are off-site and on the move.

The purpose and aim of transport security measures and procedures are to prevent transport losses and deter would-be thieves. By reducing temptation, they also help towards keeping honest employees honest. Remember that thefts often involve the co-operation of insiders! But this is not their sole effect by any means. Sensible measures and precautions can reward the prudent shipper in a number of ways.

First, the prevention of losses not only serves to *protect his property*, but if he chooses to make his own arrangements for goods-in-transit insurance, then his insurability and a good insurance record are valuable assets, too. Avoidance of losses and costly claims handling will almost certainly mean lower insurance premiums and lower operating costs. It will also lead to the imposition of fewer restrictions by insurance underwriters and therefore greater freedom of action for the shipper.

Secondly, the application of sound security thinking offers *substantial commercial benefits*. The disciplines are often no more than common sense, but they do contribute strongly to the underlying efficiency of the shipping operation. In addition, good operating practices mean good security. Organisations whose sales and shipping departments conduct careful planning and preparation receive a handsome return on the time and effort invested.

This brings us to the most vital asset of all: *customer confidence and satisfaction*. What is the point of customer service if, at the final stage, we fail to deliver or deliver short? Dependability and speed of supply are features of service to which customers give considerable weight and generous recognition, all the more since the demands of modern business dictate minimum levels of stock throughout the supply chain.

Missed deliveries damage the supplier's reputation and can cause disputes with customers that are resolved, if at all, with the greatest difficulty. Just stop to consider the huge waste of time and energy involved in troubleshooting and the scramble to make good. Will customers even allow you the opportunity and the time to attempt this? Or will they simply switch to another source of supply and turn a single instance of loss into your temporary or possibly permanent exclusion from their business? Conversely, a reliable delivery performance and the avoidance of loss and damage do reflect favourably on suppliers and can give them that crucial edge over their competitors.

RISK ASSESSMENT

The business environment dictates that all proposed security measures must not only be considered to be effective, but must also be judged to be *cost* effective. In order to be endorsed by management and to achieve acceptance and adoption at operational level in the organisation, such measures must be viewed as both sensible and practicable. Therefore, a careful analysis and assessment of the risk of loss for each type of consignment and for each routeing are desirable before the process of determining the transport method, choosing a carrier and establishing the related procedures is ever started.

It is important to be aware of the scale of vulnerability generally acknowledged for the various categories of products. In descending order of risk exposure, these are:

- Leather and fur garments.
- Other clothing.
- High cost foodstuffs.
- Tobacco products.
- Wines and spirits.
- Household electrical 'white' and 'brown' goods.
- Leather goods and footwear.
- Non-electrical household goods.
- Automotive parts.
- Non-ferrous metals.
- Building materials.

This list does not include those consignments of especially high values such as cash, foreign currency, travellers cheques, bullion and precious metals, jewellery, gemstones, works of art and antiques. The special transport methods and special operating procedures needed for these products will be discussed later in the chapter (see pp 169–70). Nor does the above list feature the most recent target of organised crime: personal computers and their microprocessor components. Within a short time the theft of this type of equipment is already reaching epidemic proportions and the determination of the thieves has taken the police and security profession by surprise. It is important, therefore, to recognise that the target of criminals' attention *does* change, just as illicit demand for various types of goods changes. Certain goods can be subject to temporary short supply, whether for commercial reasons or simply due to seasonal peaks of demand.

Additionally, the risks associated with the routeing, points of tranship-ment and the destination must be assessed carefully. For example, are we looking at Toxteth or Taunton, Baghdad or Boston? The social and crime profiles of an area are fundamental questions. Changes in risk situ-ation and risk environment must be responded to without delay. For this reason, clear lines of communication with colleagues at all levels in the organisation, with carriers, forwarding agents and official agencies exter-nally are of the greatest importance.

GUIDING PRINCIPLES

In all organisations it is either individuals or departments who initiate the need for the transportation of goods. In business this will be sales and mar-keting staff. They must be conscious of the need to involve their colleagues in Shipping or Despatch at the earliest stage of an enquiry or tender. Good planning and preparation (which means early identification of a viable routeing and dependable carriers) are critical to the efficiency and secu-rity of the shipping process. It is also important for the same reasons that sales staff give their colleagues advance notice of repeat business, espe-cially if considerable time has elapsed since the last order was delivered.

Customers should also be drawn into the planning and preparation phase and be kept informed of progress – in fact, suppliers should active-ly 'sell' to their customers the benefits of good shipping practice and sound security controls.

Forwarding agents and carriers should undergo a formal vetting process, in association with which the methods of shipment should be established, and appropriate value limits and product restrictions set for every chosen method. (A fuller consideration of this important aspect of transportation security appears later in this chapter, on pp 165–6). Close contacts should be established with agents and carriers, and these should be maintained by regular, frequent meetings. The purpose behind this is to stimulate them to thinking and working *proactively*. They can achieve this by identifying areas of potential difficulty, and by giving advance notice of changes in service and organisation. Instructions to agents and carriers (both as general procedures and when booking individual shipments) should always be confirmed in writing. Indeed, all service providers should be urged to compile a service manual which contains terms agreed with the shippers and which is available to both parties for easy reference.

A major condition imposed on the agent or carrier should be that all problems and delays are reported to the shipper immediately; a slowness to admit to difficulties may aggravate the problem and make recovery of the situation impossible. All incidents, whether these result in consignments going missing or being damaged, should be investigated without delay. Furthermore, it must be made clear to carriers that to withhold information in the hope that a problem can resolve itself will severely damage customer relations.

In the same vein, carriers should themselves receive quick internal confirmation through their own controls and systems that final delivery of a consignment has been effected. The practice of providing the shipper with a swift confirmation of delivery is fast becoming standard in the transport industry. Such service should be required of all carriers for the following reasons:

1. Non-receipt of an expected confirmation will trigger urgent response and investigation.
2. The collated data (that is, the date and time of delivery, and the date and time of confirmation) will be a valuable measure of each carrier's performance. (Recognising the potential marketing advantage, some carriers already provide monthly or quarterly performance summaries to their customers.)

Another key dimension of a carrier's service is their ability provide an audit trail of all transfer points for each consignment.

The time that goods spend at depots and hubs, at airports and docks should be kept to a minimum. Similarly, in order to avoid unnecessary periods of storage, especially at times when lower manning levels are in operation, the despatch of goods should be timed to ensure that delivery can be completed before weekends, public holidays and vacation shutdowns. Exceptions, whether by reason of service availability or commercial imperative, require special attention and monitoring.

The descriptions of freight provided by the shipper must be accurate and in compliance with the requirements of both the carrier and the international conventions that govern conventional modes of transport. However, acceptance of freight by a carrier does not mean acceptance of liability for the full value of that freight in the event of loss or damage. Shippers must be aware that the Conventions – for example, the Warsaw Convention in the case of airfreight and CMR for road transport in the European Union – do limit severely the liability of carriers. For losses over and above these limits of liability, the shipper should decide consciously whether to do nothing (and thereby in effect retain the risk himself), or to transfer that risk to insurance, either by using the carrier's special (and often costly) insurance facility, or by making his own goods-in-transit insurance arrangements under a Lloyds All Risks Marine Policy.

The security considerations relating to the documentation that accompanies the goods should be the focus of detailed attention. A central point of negotiation with carriers should be a concession to omit any sensitive wording from the description of goods that appears on the carrier's documentation, whether this is a plain consignment note, a bill of lading or an air waybill. Equally, the shipper should be careful to keep the wording on invoices and packing lists as neutral as possible, so as not to draw unnecessary and unwelcome attention to the freight. On the other hand, all accompanying documentation should include the fullest details of the consignment, that is to say:

1. the number of pieces; and
2. for each piece:
    — the mark;
    — the gross tare and set weight;
    — the dimensions;
    — the security seal numbers.

The presence of this information means that the details can be checked by the carrier(s) at the various hand-over points during a shipment.

Pre-advising customers of the full details of an impending shipment is a further essential element of transport security. Given this information and adequate notice of the shipment, customers can then make arrangements for prompt onward delivery, or for reception at their premises. They can also check the consignment details on receipt from the final carrier. Any subsequent changes in the shipping details should be notified immediately. If the pre-advice is given by fax or telex, customers should be urged in every case to acknowledge their receipt of it by sending a return message.

PACKING

The internal packing of consignments serves a dual operational purpose: safe containment of the goods and protection against damage. The outer, final layer of the packing is crucial to protecting the contents against loss and possible substitution, and also against deliberate or opportunist interference, whatever method of shipment has been used.

As a method of containment, **cartons** of substantial thickness and composition are a good starting point. *'Closed sealing'* of cartons can be achieved by fixing webbed adhesive tape to all flap edges; removal of that tape, especially the latest 'tamper-evident' tapes leaves a clear indication that interference has occurred. Plastic cross-banding around all six faces presents a further barrier to tampering and substitution.

Metal or plastic drums are a stronger form of containment, but their lid closures do not present a substantial physical barrier to opportunist interference and theft.

An effective barrier is achievable by the use of **metal or wooden boxes**. These should be of robust construction appropriate to the weight and density of the contents. Wooden boxes should be securely nailed and/or screwed, and non-return screws may even be considered for high-value consignments. Steel cross-banding around all six faces is a prudent addition, but for high values it is an essential measure to deter or at least delay thieves who may otherwise have the opportunity to access and remove the contents of the box.

Care should be taken with cartons, drums and boxes to avoid ullage or empty space, since severe jolting, whether accidental or intentional, will cause breakage and spillage. **Palletised units** should be shrink-wrapped (for cartons) and steel banded (for drums and boxes).

The shipping mark, gross weight and consignee details should be marked prominently and clearly on all pieces of freight. If goods being shipped are likely to be attractive to thieves, then it is prudent to omit the product name or description and the supplier's name from the label and packaging, unless statutory requirements make this necessary. Some suppliers, however, insist on taking every opportunity to advertise the name of their product and allow their enthusiasm for overt product promotion on their packaging to take precedence over considerations of transport security.

SECURITY SEALS

Transport security seals, if properly controlled and checked, play a crucial part in the movement of high-value goods. A broken seal directs suspicion at persons currently or just previously in control of the consignment. The design of seal should therefore only allow its removal by destruction and mutilation. Any seal, once removed, should not be capable of reinstatement. Seals are not only indicators of unauthorised entry into a container, however; they serve also to provide an indication of the continued integrity of the container and to deter opportunist tampering.

British Standard BS7480: 1992, *Specification for Security Seals*, lists the principal user requirements for security seals. They can be summarised as follows:

   The materials and construction should be able to resist satisfactorily the environments of intended use. (For example, high-strength and high-corrosion resistance is needed for sea-container seals in order

to withstand the rough handling conditions and the harsh environments at the ports and depots and at sea.)

- A seal should be easy to install correctly and easy to check for correct positive engagement.
- Seals should be easy to inspect and irregularities should be detected
readily.
- Component parts should be difficult to counterfeit or substitute,
especially in the context of high-value consignments with multi-stop
routeing over long distances.
- Each seal, other than a general-purpose one, should carry a unique
identification mark and/or a unique number for that seal type.

BS7480 classifies seals according to their strength characteristics. It considers critical design and construction aspects, specifies testing, marking
and labelling requirements, gives advice on the use of the various types
of seal and includes a code of practice for secure supply, storage, use
and recording. The new standard does commend itself, therefore, as a
valuable source of information and guidance for users when selecting
suitable sealing systems. Users should not neglect to seek the advice of
the seal manufacturers and suppliers, any of whom should be able to
draw on a wide experience of the applications and uses of each seal
type.

Barrier Seals and Indicative Seals

In general, security seals fall into two categories: physical barrier seals
and indicative seals. The function of the **barrier seal** is to present a for-
midable physical barrier to entry or access. Removal should only be possible by using substantial tools such as bolt-cutters, a saw or, in the case
of the higher strength bolt seals, an oxyacetylene torch. A barrier seal
must also provide evidence of more subtle forms of interference.

An example of an effective barrier seal is the **bolt seal**. With its formi-
dable appearance and obvious high strength, it can be expected to withstand all but deliberate attack. For this reason, it can be considered most
suitable for sealing and locking sea-containers, which are usually left
unattended at terminals and ports. When fitted, the bolt shaft should
rotate freely in the locking chamber. In addition to having this particular
feature, one of the makes of bolt seal is constructed in such a way that
the shaft only engages when it has been inserted fully through the chamber and the tip protrudes at the back. This dual feature means that deliberate incorrect fitting of the seal is easily identified at source. The
absence during a seal check of free rotation of the shaft is the tell-tale
sign that the integrity of the seal has been compromised, for the shaft will
have been fixed, most probably glued, back into the locking chamber,

thereby giving the appearance of an unbroken seal. Some bolt seal manu-facturers provide the additional feature of encasing the locking chamber and bolt head in a plastic outer moulding. Any attacks on the seal using the necessary compressive forces should leave visible signs on the plas-tic casing. In addition, if transparent plastic is used to encase the locking chamber, this will protect the structural element and prevent interfer-ence with the seal number which is stamped on the metal core.

Cable seals, although strong, are not generally regarded as providing a high level of security, not least because lax fitting often results from their complex design, and this can permit interference and reinstatement that is difficult to detect.

Indicative seals do not provide a physical barrier; they merely give an *indication* of interference. The features of a good indicative seal are:

- It is so designed that it cannot be stripped down and reassembled.
- There are no manufacturing marks, such as mechanical or welded joints, at points where they would confuse or disguise interference.
- It is manufactured from materials that cannot be manipulated or cut without leaving evidence.
- It contains identification marks, such as company logos and/or seri-al numbers, that cannot be altered without leaving evidence.

Metal strip seals have modest indicative features, but have relatively low tensile and shear strength and impact resistance.

The wire composites are stronger systems for use on drums and boxes, especially when thicker gauge wire is used. One composite with proven reliability and good indication features is the snaplock-style metal seal-on-wire system. Lead or wax seal-on-wire systems are more easily damaged, and the degradation of security numbers and marks is often a problem, due to the respective softness and brittleness of the materials.

Plastic pull-tight seals exhibit useful security features in that they can be manufactured as a one-piece moulding from plastic material that changes colour or 'blushes' when significant stress is applied and cannot be glued readily. The better quality pull-tights exhibit the feature (similar to that of the better quality bolt seals) that when fitted, there is lateral movement of the spear in the sheath. However, experience shows that even pull-tights made from the toughest nylon are not strong enough to withstand the heavy impact and chafing action to which seals on sea-containers and on drums and boxes in aircraft holds are exposed.

Seal Controls

Whichever seal is selected, it is only as good as the system into which it is introduced. In other words, to move a consignment along its routeing without conducting checks of the seal(s) at the various hand-over points is to make but a token gesture of no real security value. What is needed is a control and audit system that involves comprehensive checking and record-keeping.

The system should start with the manufacturers. Customers should satisfy themselves that a record is kept of all seals produced, to whom they were supplied, and the type of seal, the numbers and markings. The responsibility for purchasing seals should belong to one person only, and all seal orders should originate from a predetermined ordering location. Manufacturers should record for each order the date, the name of the person giving the order, the date of despatch and the name of the carrier. Seals should be supplied in complete standard packaging quantities, and delivered by a secure transport method to a named individual.

On receipt, the user should check that the seals are complete in the advised numbering sequence. He should maintain the audit trail by recording the numbers and the date of receipt in a bound seal register. The seals should be placed in physically secure conditions with access restricted to authorised staff. The issuing of seal batches to other parts of the organisation should be recorded carefully in the register. The seal register should be kept under secure control at all places of bulk storage, and regular checks of the physical stock against the register should be made.

The issuing of individual seals for every consignment should be recorded and the name of the person affixing the seal(s) entered in the register. Consignments should be sealed immediately the packing operation is complete. Drivers should not be permitted to affix seals, nor should they be allowed to remove seals at the destination points.

The numbers and marks of the seals used should be shown clearly on all documents accompanying the consignment – packing list, invoice, the carrier's consignment note, air waybill or bill of lading – so that they can be checked, confirmed and recorded:

- as the consignment leaves the supplier's premises;
- at all points of hand-over and transfer along the route;
- on arrival at the final destination.

Any replacement seals which need to be fitted during transit, whether because of damage, loss or Customs' check, should have the numbers recorded, entered on the accompanying documents and notified to all subsequent recipients. At the destination, it is important that seals are inspected for integrity and the numbers recorded. Seals should be cleanly cut, not broken, and must be retained until the contents have been checked and found to be in order.

If at any stage of a shipment irregularities are found, any broken seals and all packing should be retained until an investigation has been completed. All defective or broken seals should be disposed of in such a way that it is impossible to reuse them or any part of them.

A seal check should comprise a careful visual examination which focuses on any signs of distortion, cutting and adhesive material. This should be followed by a close examination of the seal, to ensure that it has not been violated and simply replaced loosely, or that a design feature such as free rotation of the shaft on a bolt seal is still present. In short, the purchase, storage, distribution and usage of security seals should be strictly controlled and documented, with a common set of written procedures for use within both the shipper's organisation and with forwarding agents and carriers.

VETTING AND APPRAISAL OF CARRIERS

Carriers can be defined as organisations which have responsibility for transporting shippers' property but which are outside the control of the shippers. As such, many alternative means of transport are potentially available to them, including:

- Postal services.
- Railways.
- Road haulage: groupage and full load.
- Express parcels: domestic and international.
- Forwarding agents: air- and sea-freight.
- Shipping lines: bulk and containerised freight.
- Airlines.
- Armoured-truck operators.
- Total liability carriers/international security transport specialists.

Postal services and railways are not considered here since the ability of the shipper to influence security controls is very limited.

It is prudent to establish the suitable shipping method for each type of consignment and then to set value limits or value thresholds for various levels of security procedures. It is also wise, if security controls are not to become cumbersome or operational efficiency impaired, to restrict the selected carriers and agents to a modest number. The number of carriers and agents chosen should be sufficient to ensure commercial and operational flexibility, and keen competition among those who are actually employed.

The successful companies will have undergone a formal process of security vetting and have been designated 'Approved Carriers'. The meetings and visits to premises required for initial vetting and approval should be followed up routinely by further regular contacts, the frequency of which should be determined by the security grading of the transport method and the nature of the goods carried.

A number of general factors determine the suitability of carriers:

- Their financial stability: annual reports and commercial references.
- The company ownership and history.
- The terms and conditions of carriage.
- A commitment to corporate security policy and written procedures.
- The level of security competence and the accountabilities of managers and supervising staff.
- The physical and electronic security of the premises.
- The security of information, especially when handling current shipment files.
- Computer security.
- Their losses and damage record, assembled from statistical data.
- Management skills and the quality of the training given to their staff.
- Staff turnover.
- The nature of pre-employment screening, including employees' employment history and references.
- Their policy on discipline, dismissal and prosecution.
- Other measures of performance and efficiency.
- Their achievement and adherence to quality management, including ISO 9000 series certification.
- The nature of company quality programmes and initiatives.

The level of security competence and management that can reasonably be expected depends particularly on the security grading accorded to the specific methods of transport that are offered within the carrier's range of services.

ROAD TRANSPORT

As the threat grows to goods on the road and the vehicles themselves, so is there a greater alertness on the part of the operators of conventional road transport and a growing commitment within the industry to equip vehicles with enhanced security, and to improve transportation practices and procedures. The factors to be considered when appraising the common carriers (groupage and full-load hauliers) apply similarly to the parcels companies and the international freight forwarders.

First, the **drivers:** are they in the carrier's full-time employment, or are self-employed or agency drivers used? Driving licences should have been checked and examined thoroughly for any signs of forgery before a driver is taken on. If you are a major customer, you might expect to be given the added benefit of having one driver or a group of drivers assigned routinely to your work. Because of the security implications, transport users should enquire and agree the extent to which a carrier uses subcontractors and under what conditions they operate.

Next, the **controls and disciplines to be observed by drivers**, whatever the nature of the particular haulage business involved: does the carrier employ security or other personnel to follow vehicles and check the compliance of drivers with the established procedures? Vehicles, for instance, should never be left unattended without the cab and load compartment doors being closed and locked, nor should they be parked in secluded or poorly lit locations. For overnight stops, only predetermined secure parking areas should be used; and even here it may be possible to park up to a wall or other structure so that rear or side doors cannot be accessed, thereby enhancing security. The cab doors should be locked whilst the driver is asleep.

Drivers should not give lifts to hitchhikers or other unauthorised passengers. They should never discuss their loads routes and parking arrangements with strangers; nor should they accept signals from pedestrians or other drivers other than with the utmost suspicion and caution. Vehicles and their security features should be checked for tampering and sabotage after they are left unattended, after any breakdown, and after routine servicing or repairs.

Finally, **the vehicles themselves**: panel and box vans offer greater security than soft-sided vehicles, whilst vehicles without livery and logos are often employed for higher value loads. Such vans should carry a *'vulnerable load'* card with wording approved by the Association of Chief Police Officers, for by producing this card when stopped (whatever the reason), the driver can display the instruction of his employer to proceed only to the nearest police station. Customers should consider whether their own loads are sufficiently vulnerable or valuable to require the use of groupage operators, whose trailers are driver-accompanied on ferry routes.

Box vans can be used for high-value *'full loads'*. In such cases, carriers should be able, as a matter of routine, to provide two drivers to accept additional procedures and even to provide vehicles that are installed with automatic tracking and alarm-signalling systems, over and above viable cab-fitted mobile telephone or radio equipment. For foreign destinations, the latest generation of GSM mobile telephone should be provided, since this will operate through the developing international digital networks and provide a useful backup in emergencies.

As an additional measure against bogus collections, the carrier should provide the customer in advance with vehicle and crew details, such as the vehicle make, livery, registration number and crew names, and means of identification.

The following list gives examples of security enhancements, equipment and devices which are available to vehicle operators. They are designed to delay and obstruct access to the cab and load compartment, and to protect the vehicle from being driven off. They can be specified for inclusion either from the original bodywork constructor or as a 'retro-fit' by the companies specialising in this area. Potential customers of haulage and parcels operators should look for the evidence of some or even most of these features on the vehicles used to carry their goods:

- Slamlocks on cab, load compartment and roller-shutter doors.
- High-security lock cylinder and key systems.
- Closed-hasp padlocks or bolt seals on container-style doors.
- Hydraulic cab-locks to prevent lifting of a cab unit.
- Removal or covering of side and rear glazing.
- Installation of bulkheads and 'inner skins' in panel vans.
- Siren alarm system conforming to British Standard 6803.
- Fuel and air-brake immobilisation.
- Gear and handbrake locks.
- King pin and landing leg-locks on parked trailers.

The following illustration is reproduced by kind permission of Maple, the commercial vehicle security specialists of Stockport.

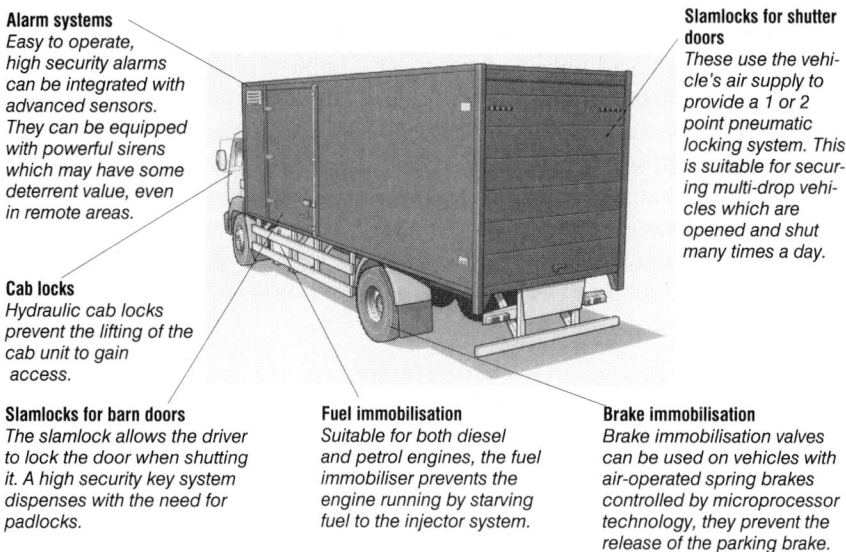

Alarm systems
Easy to operate, high security alarms can be integrated with advanced sensors. They can be equipped with powerful sirens which may have some deterrent value, even in remote areas.

Cab locks
Hydraulic cab locks prevent the lifting of the cab unit to gain access.

Slamlocks for barn doors
The slamlock allows the driver to lock the door when shutting it. A high security key system dispenses with the need for padlocks.

Fuel immobilisation
Suitable for both diesel and petrol engines, the fuel immobiliser prevents the engine running by starving fuel to the injector system.

Slamlocks for shutter doors
These use the vehicle's air supply to provide a 1 or 2 point pneumatic locking system. This is suitable for securing multi-drop vehicles which are opened and shut many times a day.

Brake immobilisation
Brake immobilisation valves can be used on vehicles with air-operated spring brakes controlled by microprocessor technology, they prevent the release of the parking brake.

Figure 9.1 *Keep the truck thieves at bay*

EXPRESS PARCELS

When assessing the security of this method of transporting goods, the control and traceability of individual consignments are crucial factors.

The sector continues to enjoy rapid growth, making it important to examine carefully the quality of the entire network – central control, distribution hubs and regional depots.

Some operators have recognised the vulnerability of certain types of consignment which customers wish to send through the express parcel systems. Indeed, they offer special services to overcome this problem. The customer should consider carefully whether to use these specific services for their own goods, in which case the question of issuing special delivery instructions – specifically, the exact location, the named individual(s) involved and the timing – should also be considered.

FORWARDING AGENTS, SHIPPING LINES AND AIRLINES

It is advisable to appoint a forwarding agent who has specialist expertise in the type of freight to be carried; one who can also offer a worldwide network of its own subsidiaries and partners. A fast 'audit trail' tracing capability is essential in the event of a loss, such a facility being more likely to be available with a specialist organisation.

It is prudent to work closely with the agent to select a limited number of shipping lines and airlines of proven reliability. For higher value airfreight, especially 'valuable cargo' as specified by the IATA, it may be wise to restrict this number still further since only airlines with high security handling and storage facilities should be considered. The chosen airlines should have a worldwide cargo network and a well-integrated security management structure.

For containerised sea-freight, the large, long-established ocean lines are reliable and likely to be able to meet the following requirements for sensitive cargoes:

- direct voyages, when available, on through bills of lading;
- consolidations on line bills of lading only;
- no transhipment, restowing or restuffing of containers without the prior consent of the shipper and active monitoring by a local agent at the port concerned.

For high-value containers, arrangements can normally be made for special stowing arrangements on board and for direct delivery and collection from the vessel as it loads and discharges.

ARMOURED-TRUCK OPERATORS

Also known commonly as Cash-in-Transit (CIT) companies, these carriers are engaged in moving the highest value loads, usually accepting responsibility and liability for the full value of the goods carried. Vetting and monitoring of these companies should be pursued rigorously, but also with a measure of sensitivity towards their need to maintain the security of their operations and to protect the confidentiality of other customers.

In most countries, armoured-truck operators are not subject to official regulation by government, except that the ability to carry firearms is either prohibited or strictly controlled. In some countries, there are security trade associations of considerable standing and prestige to which the leading armoured-truck companies belong. In the UK, for example, the British Security Industry Association (BSIA) sets rigorous minimum standards which have to be achieved by prospective CIT members and maintained by existing members. In pursuit of high standards, the BSIA issued in 1993 two guidelines to good practice:

   BSIA No 217 'Minimum Training Standards for Cash-in-Transit Personnel'.
   BSIA No 270 'Code of Practice for the Operation of Cash-in-Transit Services'.

Also within the UK, the performance-monitoring role of the Inspectorate of the Security Industry (ISI), established with Home Office backing in 1992, has won the support of those companies that are involved in the secure transportation of cash and valuables.

Specific factors to consider when assessing armoured-truck operators include:

   Are they managed professionally?
   Do they possess high levels of security expertise?
   How effective is their on-vehicle physical and electronic security?
   Do their vehicles have advanced communications, automatic tracking and alarm signalling?
   Do they have secure garaging, repair and maintenance facilities?
   Is crew morale high?
   Do they provide high-quality training programmes?
   Are there any significant endorsements or exclusions in the carrier's insurance policy?

TOTAL LIABILITY CARRIERS

A small number of the leading armoured-truck operators have each developed and linked together their national operations to provide clients with an international security shipment service. Using a network of subsidiary and partner companies, they are able to move all categories of valuable commodities virtually anywhere in the world.

The services of these total liability carriers are especially valuable to those shippers who do not themselves have the necessary shipping and security expertise or knowledge of the security transport capabilities in the country of destination. A feature of the services provided by total liability carriers is that both regular and one-off shipments can be carried out quickly and efficiently, and there is a high degree of consistency in the way that security procedures are applied.

CONCLUSION

It might be helpful to end this chapter with an indication of those factors that contribute most often to transportation insecurity. These factors have been culled from long experience of having to arrange the shipment of valuable goods to and from most countries in the world.

As a first priority, and whatever the simplicity or diversity of transport operations, it is essential to have a clearly defined policy and detailed procedures. These should be well documented in plain language and fully explained to all involved. Secondly, in all organisations, whether carriers or shippers, the managers should inspire their staff by their own strong commitment to good security practices and efficient operations. Having achieved that inspiration, goods in transit are, nevertheless, regularly targeted for theft, and the following observations are intended to help *you* to reduce the chances of becoming a victim.

Most Likely Place for a Theft

Transhipment points, eg trucker hubs, seaports, airports and sorting offices.

Key Causes of Thefts

- The labelling indicates that a package has something worth stealing.
- The accompanying paperwork indicates that there is something worth stealing.
- Poor quality packing material, which enables packages to be opened and resealed.
- Packages not secured with appropriate security seals.
- Collusion between staff handling the packages.
- Choosing an inappropriate method for shipping the goods, eg using ordinary parcel post for high-value goods.

Contributory Factors

- Those receiving the goods are not informed of an imminent/expected delivery.
- Delays in reporting losses or short deliveries.
- Goods in transit are allowed to sit in transhipment depots over weekends and public holidays.

10

DEALING WITH EMERGENCIES
Contingency Planning for Business Continuity

INTRODUCTION

Most of this book is concerned with raising people's awareness of the security dimensions of running a business by identifying threats and suggesting ways in which these can be countered. Awareness, allied with the sensible use of checks, controls, systems of work and technology, will greatly reduce the organisation's exposure to security threats. However, what happens if, despite such precautions, the organisation suffers some cataclysmic event that impacts severely on its operations – even, sometimes, to the point of threatening to end its very existence?

Before you think, *'This is all very well, but it is not likely to affect me or my organisation'*, just consider some recent widely reported events and then ask yourself if it is impossible for your business to be a victim of something similar.

- **The earthquake** at Kobe in Japan, which brought thousands of businesses to a halt and had a knock-on effect on the whole Japanese economy.
- **The bombings** at London's Canary Wharf, those in the City of London (twice), the World Trade Center in New York, and in Oklahoma City – all of which severely disrupted, over an extended period, the businesses and services that operated there.
- **The floods** in Chichester, Sussex, and through several states in the middle and south of the USA, which affected businesses every bit as much as people's homes.
- **The bush and forest fires** that almost isolated Sydney, Australia, and threatened great tracts of land in California, bringing not just

office and factory-based activities to a halt, but severely disrupting the distribution of goods and materials.

- The proposal to cut off **water supplies** one day every week to large areas of West Yorkshire in order to conserve a very scarce resource – an action which, if implemented, would certainly disrupt those industries that use water in their manufacturing or service operations.
- **The loss of a telephone exchange** through fire in North Yorkshire and two others from arson in Newcastle and Nottingham that resulted in thousands of both business and domestic telephone, fax and data transmission lines being unavailable for several days. Those firms whose computing depended on networks (both local and wide area) were especially badly affected, while many monitored intruder alarms were inoperable during this period.
- **Faults in computer software** that resulted in severe operational deficiencies; examples include the London Ambulance Service, local authority housing benefits payments, and claims-processing against the DSS Social Fund.
- **The widespread theft of silicon chips** and memory boards from personal computers located in offices, research establishments, universities, central and local government buildings. The reinstatement of the affected computers by itself represents an important drain on resources; but far more serious are those cases where vital business information has been lost as a result of the theft of computer hardware.
- **The widespread protests** in England against the export of livestock, which for a while threatened to end what was a legal business activity and added significantly to operational costs.
- The political/terrorist-inspired **assassinations** of French and other European businessmen, as well as the physical attacks on property that continue to be mounted in Algeria. Algeria has been a strong market for UK and European petrochemical and telecommunications industries.
- **The attacks** allegedly carried out by Muslim fundamentalists in Egypt against tourists, which have decimated the tourist industry and deprived the Egyptian government of vital income. Business victims include Egyptian, European and American tour and hotel operators, plus the infrastructure to support them.

These were all widely reported events, having a significance that extended far beyond the troubles experienced by any single organisation. They exclude, therefore, the routine, common-place events that can (and frequently do) impact on just one or two businesses, but which nevertheless for them create severe problems. Such scenarios could include:

- a **fire** that destroyed office, production or research buildings and equipment, whether caused by accident or arson;
- **storm or lightning damage**, resulting in local flooding of business property, power surges that disrupt computer networks, or unusable buildings, due to the deterioration of the building fabric;
- a serious **computer virus** attack;
- a **major fraud or embezzlement**;
- a prolonged **strike** by employees;
- the **extended loss of corporate computing** facilities (however this is caused);
- the unexpected and sudden **loss of key personnel** (perhaps in a road traffic accident, due to a fatal illness, or as a result of 'head-hunting' by a rival organisation);
- the **kidnap or extortion** of an employee, or his or her family;
- a **product contamination** incident (either the actuality or the threat), such as might be made against a supermarket or food manufacturer;
- a **major incident occurring in a neighbouring business**, such as the release of toxic substances, that results in your own premises having to be evacuated.

It is clear that overcoming the difficulties caused by any of these events is likely to be a priority. It is also clear that doing so will incur significant costs, some of which might be recoverable eventually through insurance. Less obvious is what impact each of these events will have on the business itself and on the staff who will be affected by it.

It seems self-evident to suggest that the continuing viability of the business will be better ensured if it develops contingency plans to deal with emergencies such as these. Perhaps, but recent research has indicated that fewer than one organisation in four in Britain has, for instance, a viable plan to recover from a major interruption to their corporate computing, only two in five have plans to cover the loss of buildings, communications and information, whilst only a quarter have decided how to recover from the loss of key staff (Hearnden, 1995). This indicates that the important issue of planning for business continuity still needs to be addressed, and it is the purpose of this chapter to examine the nature of the problem, and to suggest how best to organise an effective response to the unwanted trauma of a sudden, unexpected incident or crisis.

CONTINGENCY PLANNING

Contingency planning is an aspect of risk management with two main purposes:

1.   To plan how to recover the business from the effects of major unexpected events.
2.   To establish the means to organise the actual recovery.

One can qualify these objectives further by stating that an important consideration is to minimise the costs caused by the unexpected incident and to maximise the speed and effectiveness with which the recovery is achieved.

The diversity and range of catastrophic events given earlier should indicate that whilst crises are usually unexpected, they should not be entirely unforeseen. Human experience indicates that they do occur – some of them with annoying regularity. The human dilemma is that we are all inclined to believe that it will not happen to us! A further justification for contingency planning is that most managers have never been prepared either emotionally or intellectually to handle tragic events. They tend to equate a *crisis* with bad management and something that does not happen whilst they are in charge (Sheer, 1990).

Although it is the conventional view that disasters most commonly originate from outside the organisations they afflict, it is valuable for a moment to consider a novel dimension to the kind of corporate culture that is promoted to respond to today's rapidly changing, highly entrepreneurial, and intensively competitive global business environment. This has been identified by Dr Sally Leivesley as *the culture of danger*. Analysing the reasons for the collapse of Barings Bank, she draws attention to the wider fashion for a culture that:

> *...reinforces risk-taking with rewards – immediate gratification for the young with big bonuses orchestrated to the media in successful years...*

> *The culture of danger is a generally unrecognised phenomenon. In simple terms it is the failure of institutions to fully recognise the threat from human factors. The group accepts risk as part of daily life and rarely acts upon early cues of impending catastrophe.*

> *(When things go wrong) scapegoating of an individual, which gradually goes up the ladder, is common at the Inquiry stage, and this reinforces the cultural belief in the system being intact with a few individuals to blame...*

> *Other sectors of business and industry have had the same phenomenon – mine explosions killing teams of men have recurred again and again until the system recognised that it is not acceptable for men to die in the pits, and a whole new culture of safety has been adopted. There was the Three Mile Island incident, then Chernobyl; Flixborough, then Bhopal; Zeebrugge, then The Estonia; Barings, then whatever is to follow. All these incidents were preventable.* (Leivesley, 1995)

Contingency planning and the risk analysis that should precede it ought, therefore, to include the possibility of crises generated from within the organisation, as well as from without. Apart from the consequences of operating within this culture of risk, plans should address the possibility of major fraud or embezzlement by staff, human or system error in key activities, and deliberate sabotage.

However the crisis originates, there are the same key stages in developing an effective response. These are:

- Obtain the support and commitment of senior management.
- Identify and assess the threat.
- Analyse the impact of the various crisis scenarios on the organisation.
- Design the plan.
- Select and train those who will operate the plan (the Crisis Management Team).
- Test the plan.
- Keep the plan up-to-date.

The remainder of this chapter considers each of these aspects in turn.

MANAGEMENT SUPPORT

Although senior managers within all types of business ought to be aware of the need for recovery planning, the available evidence suggests that many avoid or overlook the issue for various reasons. Among these are misconceptions about potential threats, concerns about the costs associated with developing, testing and implementing the plan, or just the relentless nature of the everyday pressures of running a business, which eats into the time available to consider such long-term issues.

It seems likely that this attitude exists only where managers are ignorant of the true nature and extent of the threats, on the one hand, and of their potential impact on business continuity, on the other. Threat analysis and business impact analysis are in this sense essential precursors for obtaining management support, even though here they are taken as the next two stages in forming a plan for recovering from a disaster. Whatever the reason, it is vital that support is shown to exist at the top of the organisation for the contingency planning process. The reasons are the same that apply to any major policy initiative – the need to command sufficient resources, to be given appropriate priority and status within the organisation, and to target accurately all those activities that are vital to the future viability and progress of the business.

THREAT IDENTIFICATION AND ASSESSMENT

We have identified in the first part of this chapter many of the threats that are capable of causing serious business interruption. One convenient way of classifying these is to divide them into those that have their origin **external** to the organisation and those that are caused **internally.** Among the examples we have already noted, the Chichester floods represented an external threat, whilst a major fraud is most likely to be an internal problem.

Whilst this kind of classification is initially convenient, it needs further refinement before an accurate threat assessment and evaluation can be undertaken. A second reading of our list of threats reveals that they can be subdivided into those that result from **deliberate action** (the City of London bombings) and those that are **accidental** (the Kobe earthquake). These two events might well produce similar effects, but the threat assessment would be different and certainly so would the kind of countermeasures. Some people prefer to adopt a slightly different subclassification into 'natural' and 'man-made' disasters.

However, important though it is to *identify* the full range of threats that might affect the organisation, to plan an effective response requires something more. You need to know *how likely* the threat is to manifest itself, and then how severe or inconsequential *its impact* is likely to be on the organisation.

Assessing the likelihood need not be based merely on 'best guess' evaluation, for there is a great deal that can be done to reduce the element of chance involved in forecasting future events. For instance, there is a good deal of sound research (Pease and Farrell, 1993) to indicate that across a whole range of crimes the chance of becoming the victim a second time (ie having suffered once) are greater than becoming a victim for the first time. So if you are endeavouring to assess the risk of becoming a victim of, say, arson, one strong factor will be whether you have ever suffered such an attack before. If you are considering the chances of ever being the victim of a natural disaster, such as a flood, lightning strike or subsidence, there is information available from various public and private sources, such as the meteorological office, land agency, institute of surveyors, and so on.

There are, of course, common-sense precautions that you can take for yourself, like not locating your offices on the banks of a river, housing your computer suite in the basement, or below the main water-header tank. One notable example of how not to help yourself was provided by the firm that not only placed its computer suite in its basement, but appeared blissfully unaware that, on the other side of the wall, was an underground section of the River Lee. Had any disaster occurred, it would be interesting to speculate whether it should be classified as 'natural' (river flooding/undermining building foundations) or 'man-made' (culpable lack of foresight).

When the threats have been identified and evaluated in this way, the next stage in developing a contingency plan is to assess what kind of effect they will have on the organisation.

BUSINESS IMPACT ASSESSMENT

This is a crucial element for determining priorities, both in terms of deciding how much resource to allocate and which business activity to recover first after the disaster has struck. Although it makes sense to give one person the responsibility for developing a disaster recovery plan, it is impossible to over-stress how important it is that the eventual plan that is adopted by the organisation truly reflects its key business needs. This means that no single person's view of the organisation should be relied upon exclusively to assess the likely impact of specified events, or to determine the degree of priority that should be given to recovering specific activities. Instead, it is wise to obtain the views of all business unit managers about the consequences of a loss of key activities within their own areas of authority, together with an indication and rationale of how they would prioritise recovery.

The process of determining the business impact of any specified event requires answers to the following questions:

- What is the potential impact on the organisation's day-to-day business/financial capabilities of a complete and prolonged loss of this named activity?
- What would be the financial costs resulting from a loss of this specified business facility/activity for the following periods? (Usually, half a day, one day, three days, one week, two weeks, one month, etc.)
- What is the value of the goodwill and business credibility that would be lost if this specified business activity could not be provided as a consequence of a disaster?

To take just one example, the loss of the corporate computing facility responsible for processing customers' invoices might produce answers to indicate a seriously adverse effect on cash flow. (Because customers could not be invoiced, they would not be in a position to pay monies owing to the company; and because getting people to pay invoices is never easy, even when the facility was restored, the assessment might be that it would take several months to recover the backlog in payments that had built up.) The actual costs to the organisation could be calculated, based upon known cash flows, the estimated loss of revenues, and the consequent costs of borrowing to fund the shortfall. The calculation of the impact on goodwill would include not just the 'loss of face' with customers, but in all likelihood the reduced ability of the company to pay its own suppliers, and a probable deterioration of its credibility with its bankers. 'Goodwill' in such circumstances might also be taken to include the morale of its own employees, which might plummet if they were asked to work extended overtime or to forgo a pay rise because of the disaster.

The process of asking and answering questions so closely focused on the business enables one to identify those activities that are truly *critical* to its continuity. This can be further refined and priorities for recovery determined by asking the additional questions:

What would be the specific consequences of a prolonged loss of this named facility?

If the organisation experienced a major incident that resulted in the inability to carry out its normal functions, which applications would you consider most critical to recover?

Asking a variety of business unit managers their views might produce different perspectives on priorities, but it is the surest way to identify the key activities within each business sector. After this, it is right that the directors should determine overall priorities for recovery.

To summarise, then, business impact analysis is a key stage in identifying the consequences of disastrous events in a way that enables business leaders to make rational decisions about the resources needed to mitigate the effects and that helps them to prioritise business applications for recovery.

DESIGNING THE RECOVERY PLAN

To this point, the contingency planning process has been intent to identify and assess the events that pose a threat to business continuity, to determine the likely consequences and costs should disaster strike, to establish a rational basis for deciding which business activities to prioritise for recovery, and to elicit the support of senior management for the recovery plan itself.

All these factors help to form the framework for the actual recovery plan which, it has already been suggested, should be the responsibility of one senior person to develop and maintain. A prime benefit of contingency planning is that, by relating to the likely effect, rather than the causes of disasters, it reduces thinking time through anticipating problems and identifying remedial actions. Furthermore, the actions to be taken in response to an emergency have been identified rationally and calmly, away from the stress and trauma that would be present during and after the catastrophic event itself.

It is important that the plan accomplishes three things:

1. It contains measures to manage the immediate crisis.
2. It initiates actions to enable the business to continue in the short term.
3. It establishes the organisation and procedures to manage the medium- and long-term recovery.

Managing the immediate crisis is likely to be devolved to a pre-selected Crisis Management Committee (CMC), located at headquarters where all policy decisions will be made. This Committee will have agreed during the contingency planning phase on a strategy for major policy issues, which in turn will have been endorsed by the Board, some of whose members will also be members of the CMC. Suggestions for the selection and composition of the CMC are made in the next section.

A most important aspect of crisis management is establishing the means to ensure good communications and control. Efficient communications will be necessary between the CMC and the rest of the organisation, to the emergency services (police, fire, local government) and to the outside world (customers, suppliers, shareholders, the media). The best way of achieving this is to create a control room for this specific purpose, although in smaller organisations it is likely to involve just the provision of extra telephone and fax lines to an existing room. Since one of the commonest causes of disasters and emergencies is fire, some thought should be given to the location of this control room and its protection. Perhaps contingency planning should stretch to alternative locations?

The plan itself is likely to contain considerable detail of corporate operations, systems and procedures. As an example of this and to indicate the kinds of **actions that would be necessary to enable the business to resume its operations in the shortest possible time**, here are just two aspects of the planning to cope with the loss of head office in a fire. For a start, it is going to save a lot of time reprinting stationery (invoices, letterheads, despatch notes, orders) if the document specifications are available, together with artwork and the names, addresses, telephone and fax numbers of several printers. Again, if the fire destroys 20 Personal Computers, the plan should describe how the data can be recovered, what versions of what software are used and from where immediate replacements can be obtained, what arrangements have been made to replace the actual PCs, who to contact to re-establish the local area network, and what alternative accommodation has been arranged to house the staff, together with any special transport arrangements that might be needed.

Already, two things should have become clear about contingency planning: it requires meticulous attention to detail if it is to cover all foreseen eventualities, and it must be *dynamic,* in so far as it must constantly reflect where the organisation is now, not where it was two years or even six months ago. **Keeping the plan up to date** is therefore likely to be a full-time job for one person in a large organisation, and an important part of someone's work in a smaller business. Good contingency planning, therefore, does incur a cost. The benefit (more accurately, the cost-benefit ratio) is a judgement for senior management. What would be a mistake is to ignore the potential for catastrophic events within the increasingly technological environment in which businesses operate today.

The final element in any contingency plan is to lay the foundations for **the long-term recovery and re-establishment of the business**. Although staff are likely to display something of *the 'Dunkirk spirit'* in their initial efforts to recover the business – for example, by working in temporary accommodation, perhaps a long way from home, and possibly working unsocial hours – this is likely to fade over time, making it important to implement more permanent measures and to return to normality.

THE CRISIS MANAGEMENT TEAM

The Crisis Management Committee will typically comprise the following members, although it might wish to co-opt other specialists, depending on the nature of the problem facing it.

- **Chairman.** The Chief Executive (or his nominee), who can authorise actions and expenditure.
- **Deputy Chairman.** A senior manager, fully acquainted with the contingency plan, who can assume some of the workload of the Chairman, and act in his place if he is a victim of the crisis.
- **Crisis Co-ordinator.** A senior manager who acts as executive secretary to the CMC.
- **Financial Adviser.** Someone who has detailed knowledge of corporate finances, can liaise with the banks, raise additional funds when required, manage cash flows.
- **Human Resources Specialist.** To advise on employment matters, deal with contract agencies, seek to maintain staff morale, deal with the families of any staff victims.
- **Legal Adviser.** An important member of the committee, probably drawn from the organisation's legal advisers, to provide expert guidance on all those matters which have a legal dimension.
- **Security Manager/Adviser.** This person might act as the crisis co-ordinator, but would in any case be responsible for the security of meetings, communications and the recording of committee decisions (especially where these carry legal implications).
- **Public Relations Adviser.** Someone versed in dealing with the media, whose task will be to promote a good corporate image in all dealings with the press, shareholders, customers and suppliers.

For some types of emergency, for example a kidnap, an extortion or a product contamination incident, the committee will need **a negotiator**. In the UK, the police will assume control of these kinds of incidents, once they are reported to them, but there are nevertheless many occasions when someone from the victim organisation will become directly involved in this capacity. Clearly, the negotiator should be experienced in the role, because the results of a mistake are likely to be extremely serious. For this reason, most of the companies providing insurance against such incidents will have an arrangement with one or more firms of consultants who specialise in this area.

TESTING THE PLAN

We have already made the point that any contingency plan needs to be dynamic, reflecting the current systems and organisation it seeks to protect, not those that applied some time ago. This clearly indicates a need to monitor and update the plan constantly. But how can one be sure that the plan will achieve what it sets out to do, that the arrangements made for recovering from the disaster will actually work, and that nothing significant has been omitted from the plan?

The answer is to test the plan by simulating the situation(s) to which it is meant to apply and invoking the recovery plan. With complex plans, it may be advisable to test one part at a time, rather than the whole plan. There is little doubt that an exercise along these lines can be expensive to run, but it is common experience that contingency plans are found to be faulty in one or more aspects when they are first put to the test. A survey of 39 computer mainframe sites, for example, found that only two-thirds had recovery plans, of which only half had been tested. Worryingly, significantly more than half of these failed during the test.

So if testing is the only sure way to establish the quality of the contingency plan, but is known to be quite costly, is there any halfway house that would provide some measure of reassurance? The most sensible compromise, if costs are a problem, is to select just those scenarios that are considered to present the highest risk – for example, the loss of corporate computing facilities – and to test the effectiveness of that part of the contingency plan that relates to it. If even this is considered to be too disruptive, testing might be restricted to one department at a time, by metaphorically pulling the plug on their part of the computing service and seeing how well they recover their own operations. A successful invocation of the plan can be a very positive stimulus to staff morale.

Any errors and omissions revealed during testing will need to be incorporated in an updated version of the plan, reinforcing the dynamic cycle of develop, revise, test, revise, develop, and so on. On the question of testing, though, remember that what we are considering here is another (very practical and hopefully effective) form of insurance, and the last thing one wants is an insurance policy that fails to pay out in an emergency.

CONCLUSION

Contingency planning is a sufficiently common aspect of management to cause a few raised eyebrows when it is advocated as a valuable tool within the security portfolio. However, what we have just explored are the specific ways in which this important discipline can be harnessed to support the key activities of crisis management and business recovery. In this role, it is also associated with risk assessment and evaluation, which are the necessary precursors that allow businesses some way along the path to managing crises proactively, rather than merely reacting to them once they occur.

The problem is expressed quite forcibly in the *Handbook of Security* (Kluwer):

> *In an ideal world perfection would be the order of the day. Complete mastery of our planet would enable us to identify and eliminate all threats to life and property and to create an environment which was entirely benevolent. However, our experience to date indicates that we have still a long way to go in order to attain this idyllic state of affairs. Meanwhile, we have to admit failure, whether it is brought about by natural causes, accidentally or through crime. The size, concentration and complexity of 20th century technology increase the probability that a minor failure in the system will produce adverse repercussions on a large scale...*

> *It is sometimes argued that disasters never occur in a way or at a time which can be foreseen, and that consequently there is no point in planning for them. This is tantamount to relying on 'breakdown' as opposed to 'preventive' maintenance, and will prove to be a similarly short-sighted and expensive policy.*

References

Hearnden, Keith, 'Corporate Computing: Key Business Issues', in *Computer Audit Update,* Elsevier Advanced Technology, Oxford, May, June, July and August 1995.

Kluwer, 'Disaster Planning', Section 3.7 in *Handbook of Security* (1st edn), Croner Publications, Kingston upon Thames, London.

Leivesley, Dr Sally, 'The Culture of Danger – Barings Collapse' in *Safety Audit,* Purley, London, April 1995.

Pease, K and Farrell, G (1993) 'Once Bitten, Twice Bitten: Repeat Victimisation and its Implications for Crime Prevention', Home Office Crime Prevention Paper No 46, HMSO, London.

Sheer, T 'Crisis Management', in *The Principles of Risk Management* by R Jenkins and K Aiken, course notes to Risk Management Module at Loughborough University, Tillinghast, London, 1990.

11

TERRORISM

TERMS OF REFERENCE

We tend to think of terrorism as a late twentieth-century phenomenon, whereas in fact it has a long history. Certainly, there have been violent actions in pursuit of political aims for centuries: it is just that the perspective of history gives a different gloss and often adds a legitimacy that was not so evident at the time. Wat Tyler and the peasants' revolt in the fourteenth century, the Luddites in the nineteenth, the French Revolution, the assassination of Archduke Ferdinand in Sarajevo, that triggered the First World War, the revolt against English taxes that resulted in tons of tea being thrown into Boston harbour were all violent actions against the establishment.

Nevertheless, being able to cite historical precedent in no way diminishes the very real threat of extremely unpleasant and traumatic events overtaking the businessman at some time, either in the UK or abroad. For the purposes of this chapter we shall include within the broad term 'terrorism' the following events:

- Bombs
- Kidnaps
- Extortion
- Sabotage
- Product Contamination
- Hijacks

A profile of the risks will be followed by a review of the precautions that can be taken to reduce the chances of you or your organisation becoming a victim and, if the worst should happen, of being able to recover with least disruption.

BOMBS

Explosions caused by bombs are perhaps the best known and most feared of terrorist actions. In the UK we have experienced for many years the violence of bombs detonated by the IRA and similar groups in pursuit of a political agenda in Northern Ireland. Although most of these have exploded in Northern Ireland itself, there have also been several extremely damaging events on the British mainland – two in the City of London, in Manchester, Salford, Tyneside, Leicester, at Canary Wharf in London, and so on. Several other bombs have been detected and defused before exploding, often causing immense disruption to normal life and to trade – the bomb placed under a bridge on the M6 being a conspicuous example.

But recent problems of this kind have by no means been restricted to the UK. Algeria, for example, has been going through a most turbulent period, when many hundreds of civilians have died. The USA has suffered three of the most damaging recent bombings, at the World Trade Center in New York, in Oklahoma City and in Nairobi, Kenya. France has also had its share, with explosions in their Metro system the result of protests by the Algerian Armed Islamic Group (GIA) against the capture, imprisonment and trial of an Arab terrorist. Moving outside Europe, there have been recent bombings in Cape Town, South Africa; New Delhi, India; Georgia; Israel; and Brazil.

In so many of these and similar bombings, businesses – rather than the state itself, the object of the terrorists' ire – have been the principal victims. This is not always by accident, since terrorist organisations have long recognised the political pressure that results from targeting businesses, as well as the extensive publicity it generates. Take just the World Trade Center and the Manchester shopping centre explosions as examples, and you reveal many examples of small, single outlet traders having their sole source of business taken away from them for an extended period, sometimes resulting in the closure of that business. Even for larger organisations, the impact will have been seriously disruptive and largely unmitigated by insurance, since these are events for which it is very difficult to arrange cover. In such situations, there is much pressure on the government of the day to 'do something about it' and to provide some sort of financial recompense for those affected, whilst the events themselves obtained a great deal of media exposure at the time and afterwards.

It seems no more than pragmatic, then, for businesses to recognise that they may well become the innocent victims of someone else's argument, and need to reflect this possibility in their business plans. To identify this possibility is an essential first step in assessing and managing the true nature of the risk, at which point the process of developing a response to the threat of terrorism (however this should manifest itself) can be integrated with other aspects of business risk management. What chance is there of the event occurring? What will be the impact on the business? What contingency plans should be developed to mitigate the impact?

Are there any general clues about potentially dangerous situations for businesses? It is clear that there are, though they often amount to no more than common sense. A good indicator is the presence of political tension in the territory where the business is operating. Thus, the conflict between the nationalist movement in Northern Ireland and the UK government was well known, and the consequent general risk from IRA actions readily recognised. Converting this general awareness of a risk into one specific to a particular business is much more difficult. There are some general clues, such as (in the IRA example) having a close association with the UK government (for instance, through being a defence contractor); like being a flagship enterprise (as in a national airline or dominant petrochemical company); such as making a major financial contribution to the political party of government; by operating as a significant player in a key national economic activity (thus, the attacks on clearing banks, the stock exchange, Lloyds and insurance companies in the City of London); or by being physically located close to a government building, (which may well itself be targeted).

Moving away from the specific tension arising from the troubles in Northern Ireland, there are other general indicators of enhanced risk from bombings. One that has an unfortunately long history associated with it is being an American company operating abroad. As the world's most powerful country, the USA has had thrust upon it a world-wide leadership role, which inevitably brings it into conflict with groups whose interests run contrary to theirs. There also exists the resentment of power and wealth among those who do not share them. The consequences have often been that American companies located in foreign countries have been attacked just because they are American. For a while in the 1970s and early 1980s, there were a spate of bomb and arson attacks on the computers used by American companies in Europe: nowadays the victims tend to be the national flag carriers, like oil companies, airlines and banks.

Currently, the USA is still reaping the whirlwind resulting from its conflict with Iraq and its wider alienation of the Islamic world – a problem that Britain, with its close support for American policy, may yet also inherit. An indication of the potential for violence came in a statement on 22 August 1998 from the International Islamic Front (IIF). In faxes sent to the London base of the fundamentalist group al-Muhajiroun it said that the IIF was mobilising against America and Israel to bring down their aircraft, prevent the safe passage of their ships, orchestrate occupation of their embassies and force closure of American and Israeli banks and businesses. Two days later, Sheikh Omar Bakri Mohammed, described as the British spokesman of the IIF, said that the possibility of attacks in Britain and Europe should be taken very seriously, and that he would expect Muslims in Europe to take action.

For strictly commercial reasons, there are many pressures on businesses to trade increasingly widely throughout the world: but this inevitably exposes them to operating in countries that can be dangerous. The problems that might be encountered are not necessarily bombs, but threats to personnel, hostage taking, extortion and sabotage of plant and operations (of which more later). The current tensions in Russia and south-east Asia brought about by financial crises have added to the problems of managing business risks there, as law and order buckle under the strain of placating wide swathes of the population, who are the immediate victims of the economic turmoil.

Some parts of Africa and South America offer similar problems, making it highly desirable for organisations wishing to operate in such areas to obtain specialist advice and information about local conditions. Information is usually available from embassies abroad and from the Foreign Office, but is also obtainable as a commercial service from organisations like Control Risks Group and Pinkertons Global Intelligence Services. This normally operates on a subscription basis, in return for which regular briefings, incident reports and world-wide risk assessments are provided to subscribers.

It should be stressed that this kind of service does not restrict itself to bomb attacks, but seeks to cover all those events likely to affect adversely business operations in the designated territories. To understand a little more about what these are, we shall now move on to look at some of them.

KIDNAPS AND EXTORTION

It was not so long ago that kidnaps were a regular event in Italy, where the motives were more often criminal gain than political pressure. Typically, a member of a rich person's family would be taken, with release dependent upon payment of a sizeable ransom. Sometimes, the victim would be an employee of a prominent business, where demands for a ransom payment were difficult to resist without incurring public opprobrium. Similar events occurred in Britain, but with less regularity. The reasons for this were not absolutely clear, but might have been connected with a perception that the British police dealt with such incidents more reliably than their counterparts in Italy. Certainly, the British police invariably managed to obtain the co-operation of the media in keeping details of such incidents out of the public domain until they were over, thereby denying the criminals access to the kind of pressure brought about by public debate and the scrutiny of crucial decisions. A useful by-product of this policy has been a lessening of the likelihood of 'copycat' crimes, another factor that has contributed to the relatively low numbers (around 70 to 100 a year) of recorded kidnaps in the UK.

More recently, kidnapping has been a significant problem for businesses operating in South America, with Peru being accorded the dubious distinction of 'Kidnap Capital of the World'. Here the 'Shining Path' guerrillas have for some years kidnapped foreign businessmen, using the ransoms to fund their struggle against the government. Colombia, afflicted by the tensions induced by the drug cartels, has also witnessed many kidnappings and killings.

Further north in Mexico a man has just been arrested who has allegedly been personally responsible for fifteen kidnappings. Also in Mexico, the Jalisco state prosecutors have accused several police officers of being in league with kidnappers and have already detained two policemen. Earlier in the year there were accusations that the head of an elite anti-kidnapping unit was himself a major kidnapper. (Parts of Mexico have offered a rather unstable environment recently, as may be gathered by a report that the family of an American citizen who disappeared in the border city of Ciudad Juarez has asked the US government to help find him. They claimed that 199 people had been reported missing from this one city since 1994, of whom 19 were US citizens. All of those who disappeared are alleged to have been involved in drug trafficking or some other illegal activity.)

Indication of the world-wide nature of the problem comes with another report from Nigeria, where an unspecified number of Texaco employees were held hostage in August 1998 in the oil-producing region.

Back in the UK, a recent development has been the advent of 'tiger kidnaps'. These differ from others primarily in their short duration – the conventional kidnap often results in the victim being held for many days, sometimes weeks, and occasionally months. Tiger kidnaps are almost exclusively directed against employees of businesses, and involve extortion. Typically, an executive will be targeted who has access to cash or valuables – such as the manager of a bank or store, or one who has personal wealth. One or more members of his close family group will be held hostage, and their safety used as a lever to persuade the executive to pay a ransom. Again typically, the executive is only allowed a very short time to obtain the ransom, thereby leaving a much reduced possibility of organising a thoughtful, planned response. These are criminal, rather than terrorist actions: but the distinction is of no interest to the victims.

In another example of extortion, the terrorist/criminal will threaten a business with a bomb unless a ransom is paid. In Britain both the Sainsbury Group and Barclays Bank were blackmailed over a two to three year period by the 'Mardi Gra' bomber, who exploded a number of relatively small bombs on or immediately outside their premises. Both organisations were thought to have involved the police, who mounted an extensive operation before eventually catching the two men allegedly responsible.

SABOTAGE

Exports totalling about 800,000 barrels per day, or just under half of Nigeria's total crude oil output, have been affected by two separate incidents, according to Shell officials. Force majeure at Bonny was declared on Saturday after a main pipeline linking inland processing plants with the coastal export terminal was sabotaged.

Investigators found that a 9 mm hole had been drilled in a buried pipeline, causing an estimated 700 barrel spill. The operations of Shell and other joint ventures between foreign oil companies and the state-owned Nigeria National Petroleum Company in the Niger Delta are routinely interrupted by what the oil companies call 'community actions'. These range from sabotage to the temporary occupation of production facilities. (*The Financial Times*, 25 August 1998)

In Brazil a bomb destroyed an electricity pylon and thereby caused widespread power cuts. (September 1998)

In Britain and Ireland the IRA regularly targeted commercial premises with bombs, while in Spain the Basque separatist organisation, ETA, has pursued a similar policy.

In the UK, the Animal Liberation Front (ALF) has for many years attacked research laboratories where animals were used in experimentation. In August and September 1998 it staged three raids on mink farms to release mink into the wild. Apart from the commercial cost, another consequence was likely to be widespread environmental damage, since mink are aggressive predators and represent a significant threat to both domestic animals and other wildlife.

Given that one frequently held objective of terrorism is to attack the state by causing economic disruption and loss, it should come as no surprise that industrial and commercial sabotage is a frequent weapon. Here are some examples.

PRODUCT CONTAMINATION

There are often close links between product contamination and two other phenomena we have already discussed – extortion and sabotage. The methodology is to contaminate, or threaten to do so, a product in such a way that the consequences to anyone subsequently using that product would be disastrous. Usually, it is a food item that is chosen, with poison, broken glass, a pollutant or a corrosive agent added surreptitiously. The perpetrator relies on the difficulty of the victim organisation detecting the contaminated product, the high cost of organising a product recall and de-stocking of (say) supermarket shelves, and the extremely adverse publicity that would accrue to the victim organisation should one of their customers be injured or killed by the contaminated product.

It can readily be seen that the organisation targeted by the terrorist/criminal could be either a food manufacturer or the store owner. Examples exist of both. Some years ago in the UK, Heinz baby foods were contaminated with powdered glass, whilst in the US and Japan brand leader, Thylemol, was the victim of a prolonged campaign.

There have been recent examples again in Japan, starting with a mid-1998 contamination of food at a festival in Wakayama, as a result of which four people died and more than fifty were made ill. In September of the same year, in one of three separate incidents a man was poisoned by cyanide that had been inserted into a can of soda bought in a supermarket. In the second, a supermarket employee was arrested for putting insecticide into a thermos of hot water and some salad dressing in his firm's cafeteria – his stated motive was dissatisfaction with his job. The final case involved a carton of coffee laced with a lethal dose of sodium cyanide, which was fortunately detected before being consumed, because it gave off a strange smell. Whilst it can be argued that the Japanese cases were likely to have been copycat crimes, and not the result of concerted action by one individual or group, the actual impact is much the same, whether attributed to individual 'criminals' or 'terrorists'.

Although these examples all relate to food products, there are few reasons why the technique could not be applied in other spheres – to engineered products, especially consumer durables, for example. (Imagine the problems if car brake discs were somehow corrupted to fail early, perhaps by a disaffected employee, or gearboxes similarly attacked, or incendiary devices wired into the backs of television sets!) Something very similar to product contamination has already been used in IT, where software code has been deliberately corrupted to cause failure and disruption. Here the hoped-for payoff occurs when the perpetrator is hired to solve the problem s/he has created, or is paid a ransom for revealing the exact cause of the problem.

To summarise: product contamination has usually been perpetrated by lone criminals, rather than a group of terrorists. However, the impact on the target victim of a successful campaign can be just as catastrophic as the effects of a bomb. The cost to Heinz, in the case cited earlier, was estimated to be many millions of pounds in terms of taking their baby food lines off the supermarket shelves, then recalling them, losing sales initially, and then seeking to recover customers' confidence. In the case of Thylemol, a world-wide brand leader was reduced to a non-runner over the many months of what was later considered to be a poorly handled crisis.

HIJACKS

Concerted action by the governments of the world and by the International Air Traffic Association (IATA) has eventually succeeded in reducing the number of aircraft hijacks, which in the 1970s and 1980s were a prime outlet for terrorist protest. As already mentioned, airlines then (rather more than now) fulfilled the role of national flag carriers, so that they were particularly vulnerable to attack by extremist groups with a grudge against the state they represented. American and Israeli planes were the most frequently targeted, but El Al swiftly imposed very tough security measures to protect themselves – and to a large extent still do to this day.

Most of the hijackings were undertaken by Arab groups, though there were at one time many incidents within the People's Republic of China. During the height of the tension between the PRC and Taiwan, word spread that the Taiwanese government would pay a king's ransom to anyone who brought a PRC jet across to Taiwan. Unfortunately, these payments were for Chinese military jets: but it was reported that a number of rather second-hand BAC 1-11s were diverted at gunpoint across the Straits of Formosa by various Chinese peasants following the path to instant riches.

Although aircraft hijackings have thankfully declined, there remains more than ever a general issue of travel safety which business executives need to address. There are many countries in the world where journeys to and from airports, and movements within and between cities are far from safe, whilst it is no more than common sense to consider carefully the choice of airline and destination at times of particular international tension. Choosing an American airline, for example, to fly to the Middle East in the immediate aftermath of the USA's bombings of targets in Afghanistan and Sudan in August 1998 would not have represented the safest choice, were there an alternative flight available from an airline operated by a 'neutral' country.

General travel risks include those that we have examined in this chapter – kidnap and extortion, becoming caught up in local unrest, in which gunfire and bombings might figure and, of course, theft. Most large companies have developed travel safety advice for their executives, along with contingency plans for dealing with any emergencies that should arise. For those organisations that have not yet done so, advice is available from a number of sources – from Pinkertons and Control Risks Group in the form of subscriber information and assessment services, or from specialist security advisers, whom reputable associations like the Association of Security Consultants and the American Society for Industrial Security will be able to recommend. Some advice on travelling safely is given in the next pages.

MANAGING THE RISK

Organising an effective, business-oriented response to the threat of terrorism is a prime example of risk management methodology. The principal elements will now be reviewed to conclude this chapter. Chapter 2 has already provided an overview of risk management, while Chapter 10 examined the problem of dealing with emergencies. The latter chapter, in particular, has great relevance to the threats from terrorism and should be borne in mind whenever this problem is being considered.

Like other threats to security, the risk from terrorism needs to be evaluated at the outset. This evaluation will seek to determine the probability of a terrorist event impacting on your organisation and will then endeavour to estimate the cost should it happen. If your organisation employs its own security adviser, then evaluation of the risk should properly be assigned to him or her: if not, then it is important that a senior executive is given the responsibility. This person might well seek help by obtaining advice from the police; from the Foreign Office (if the evaluation is being undertaken for operations abroad); by subscribing to a risk assessment service; or by engaging a reputable security consultant.

Businesses can help themselves to a significant extent by being politically aware of what is happening in all those territories where they have operations. There is a great deal of relevant information available from the quality newspapers in the UK, from some trade associations and from news abstracting services, quite apart from those sources just mentioned. Do not just ignore the problem, because without specific delegation of responsibility the risk will not receive due consideration.

It is quite legitimate – after proper consideration and evaluation of the risk - to decide to retain the risk, if the consequences to the business of a terrorist event are considered to be small, or the event so unlikely as to make the cost of countermeasures unjustifiable. Risk retention is one of four possible responses. The others are:

- risk avoidance – whereby you act to remove that part of the organisation's operations at risk from a terrorist act, for example, by ceasing to send employees to areas of high risk;
- risk transfer – typically, through insuring against a potential loss, but also possibly by subcontracting the operation at risk. Thus, it might be possible to sub-contract the breeding and housing of animals to be used in laboratory experimentation, in order to avoid the risk of attack on your own organisation by the Animal Liberation Front;
- risk reduction – through safety and security measures that reduce the likelihood of a terrorist action succeeding. Typically, these will involve physical security measures (target hardening), improved procedures and controls, and good intelligence gathering and analysis. Physical security measures might well include security film on

building windows, since experience of many explosions shows that damage from glass shards is a major source of deaths, injuries and property damage. Improved procedures will address the important issue of building evacuation in the event of a bomb threat, and the provision of secure shelter for staff. Risk reduction measures must also include plans to recover from any disaster, so that the impact on the business is minimised.

Business impact assessment and contingency planning for disaster recovery are key elements in responding to terrorist acts, just as they are fundamental to any security strategy that is wholly focused on the needs of the business (see Chapter 10). Part of this process is likely to involve the establishment of a crisis management committee and procedures for regularly reviewing and updating contingency plans. In the specific case of terrorist threats, planning and crisis management are likely to be complicated where they must embrace operations abroad, and where there is a need to ensure the safety of UK-based staff travelling to and within countries that do not always offer the developed infrastructure and relatively stable environment we are used to here.

Although travel safety should ideally be tailored to the local conditions and based upon specific knowledge of local risks, there is nevertheless some sensible general advice that will reduce the chance of the traveller becoming a victim of crime. The tips that follow are not in any order of priority, but when combined in practice take out the most common opportunities for criminal attack on travellers:

- Use major airlines where possible, since these operate better security procedures, and also maintain their aircraft to a high standard, thereby minimising the chances of mid-air crises, delays and diversions.
- Arrange to be met at the airport

or

- Use hotel courtesy buses or a *pre-arranged* taxi service.
- Do not frequent red light districts.
- Do not hail street taxis – ask the hotel or the organisation you are visiting to arrange one.
- Dress in a style that blends in with the locals.
- Always let someone know what your telephone number is, and check in with them regularly. (Your failure to make an anticipated call is a good early warning of possible problems.)
- Carry a mobile phone, and memorise local numbers that can provide emergency assistance.
- Do not wear expensive jewellery, watches, etc.
- Take travellers' cheques, not large amounts of cash.
- Do not use street ATMs – go into a bank, or use the hotel facilities.

In summary, terrorism is unlikely to offer a major threat to all businesses: but especially for those trading abroad it is one factor they should include in their assessment of business risks. Political situations that give rise to conflict and sometimes spawn 'terrorism' are constantly changing, making awareness and sensitivity to these issues an important part of a balanced and informed response. Good intelligence, however it is obtained, is therefore a particularly valuable tool in this context, and ignorance is most definitely not bliss.

12

DON'T BECOME ANOTHER VICTIM
Frauds and Scams of External Origin

Although it is wise to look first to employees as the source of frauds and thefts, there are, nevertheless, a number of scams which are inflicted on companies from outside. Some of these have been around for a long time, but would they have survived if they had not achieved a measure of success? Brief descriptions follow. Can you be certain that none of these scams have been successfully inflicted on your own organisation?

LONG FIRM FRAUD

This fraud operates by another 'business' obtaining credit for the supply of goods from you and then literally disappearing before the debt can be collected. The fraud usually starts with a request for supplies, based upon the provision of trade and sometimes bank references, and is accompanied by a ready agreement to show good faith by making *pro forma* payments for the first few transactions. It is common business practice to grant credit facilities after establishing a business relationship in this way; but it is then that 'the sting' is made. A further, probably larger, order is placed for goods, which are this time delivered on credit. The invoice is not paid by the due date, whereupon most credit controllers would initiate recovery action. However, by this time the long firm fraudsters have disappeared from their known address, which was probably taken on a short-term lease, leaving no trace of their whereabouts and very little chance of ever recovering either the goods or the money that is due.

There are some useful checks you can make to minimise your exposure to this kind of risk:

　　Be wary of recent changes in the management of the new customer.
　　Be suspicious of any business that gives only a PO Box number as its address for communications.

Likewise, be suspicious of any business that uses an answering service.

Check trade references thoroughly, and look out both for those who cannot be contacted and those who give an immediate good report, without looking at records. The former may not exist, whilst the latter may be parties to the fraud.

Flag credit balances that begin to increase significantly.

Beware of new orders much larger than those previously received, especially when these are marked 'Urgent'.

THEFT OF PROPRIETARY INFORMATION

We have already identified (in Chapters 6 and 8) many sources of external threat to the security of a company's confidential information, some of which operate in conjunction with employees of the victim organisation.

The methodology of the theft will vary widely, but the most likely variations are:

Theft of the **PC disk** on which the desired information is stored.

Bribing an employee to steal the information required.

Placing an employee or contractor within the victim organisation. Do not overlook the opportunities that are available to people like office cleaners, who usually have unrestricted access to the premises at times when others are not working. Why shouldn't a cleaner be computer literate, for example?

Scavenging in dustbins for discarded corporate documents, such as drafts of board papers, spoilt or surplus photocopies. (The cleaner might already be doing this, anyway.)

Eavesdropping on careless talk by managers in public places.

By listening to **conference presentations**. (This especially applies to research staff who gain status among their peer group by giving papers at such events. However, there are professional associations for accountants, marketing managers, engineers, computer professionals and many more, most of which run regular conferences at which your staff might be asked to speak.)

Electronic surveillance ('bugging') does happen, and there is a wide range of different equipment from which to choose. However, when there are so many simpler alternatives, the risk is likely to be small, unless your organisation is going to be specifically targeted because of your involvement in a takeover, a severe industrial dispute, or because of the sensitive nature of the work you undertake. If it does concern you, there are reputable specialists who will

'sweep' your premises for bugs, of which one of the best known is Audiotel (see Chapter 13).
- **Hacking** into your computer database is currently more of a threat in the USA than in the UK. However, if your computer systems are accessible from the Internet or have dial-up facilities, then potentially you are at risk. This particular threat is covered more fully in Chapter 8.
- Other companies' **representatives and maintenance personnel** often legitimately visit not only your own premises, but those of your rivals, suppliers and customers. Careless talk in their presence could be repeated, and possibly distorted, elsewhere, to the detriment of your organisation.

THE NIGERIAN FUND TRANSFER SCHEME

There is no intention here to be racist in any way, but the fact remains that many individuals and companies have received letters offering access to sizeable sums of money, acquired in rather dubious circumstances, in return for the targeted victim's co-operation – and all those that I have seen originate in Nigeria. The fraud (for that is what it is) seems to be based upon the premise that there are sufficient people in the world who are so greedy for a share of someone else's wealth that they are willing to participate in a scheme of patently dubious morality to secure it.

The scheme starts with the receipt of an unsolicited letter (see the example opposite) describing how the writer and his associates have managed to acquire a large sum of money from over-priced contracts, or in some other way. Credibility is sought by linking these funds to a Nigerian state institution or a corporation of international status (like the Nigerian National Petroleum Company).

DR. HASSAN MUHAMMED
TEL/FAX: 234-1-4528447
LAGOS - NIGERIA.
27th June, 1995.

Dear Sir,

REQUEST FOR URGENT BUSINESS RELATIONSHIP

My request is anchored on our strong desire to establish a lasting business relationship with your company. You have been recommended by an associate who assured me in confidence of your ability and reliability to prosecute a pending business transaction of great magnitude requiring maximum confidence.

My colleagues and I in the Nigerian National Petroleum Corporation (NNPC) are looking for an Overseas firm that is trustworthy and reliable whose account can receive a huge sum of **US$28,600,000.00 (Twenty - Eight Million, Six Hundred Thousand U.S. Dollars)** resulting from a back - log debt of an over-invoiced contract at the moment in a secured suspense account of the Central Bank of Nigeria (CBN). I am to explain and arrange with you the possibility of transferring this fund from the Central Bank of Nigeria to your Company's account or any other account nominated by you. This outstanding debt has to be applied for by a Foreign contractor and payment can only be made into a Foreign account hence this contact is necessary to accomplish the deal.

In the arrangement, your firm is to act as a subsidiary to the original firm that actually executed the contract. We shall therefore need your company Letter-Head, Invoice and bank particulars which we shall use for the application and subsequent payment. Be rest assured that all modalities have been meticulously put in place for a successful transfer.

If the deal interests you, please indicate your interest through the **tel/fax number 234-1-4528447** by sending these documents and bank particulars. We will share the funds in a ratio which will be satisfactory to all concerned under Agreement, however, a 60% for us (the officials), 30% for the account owner (you) and 10% of the sum involved shall be used to pay back all expenses incurred by either parties during the course of the transfer of the funds before sharing.

This business should be kept secret and confidential because my colleagues are top government officials who would not want to experience any kind of exposure. I wish to assure you that once the transfer is initiated, it is expected to take 10 working days for the funds to get into your nominated account.

Please be informed that my colleagues and myself will be depending on your kind advise and directives as we wish to privately invest our share of the funds into any viable business as may be advised by you in your country. Please state your private telephone and fax numbers for fast and easy communication.

I solicit for your intimate co-operation and early response in this business proposal. The detailed plan will be furnished to you on receipt of your response.

Best Regards.

DR. HASSAN MUHAMMED

The problem the writer and his friends have to overcome, and the reason for soliciting the victim's assistance, is the difficulty of transferring money out of Nigeria. To achieve this, the writer asks the victim to provide him with certain documentation, in return for which he offers a percentage (usually around 30 per cent) of the monies to be transferred to the UK. The documentation required is the victim's:

- Name.
- Telephone and fax numbers.
- Bank account number.
- Banker's name, address and telex number.
- Company (or personal) letterheads.
- *Pro forma* invoice.

It is probably clear already how the basis for the fraud is now established, but to illustrate it fully, here are the three ways in which events might proceed. First, the victim's documents can be used to extract money from his account – after all, the fraudster has possession of all the necessary details to request the bank to transfer money to *Dr Hassan Muhammed* (in our example). Alternatively, the victim could be invited out to Nigeria on the pretext that the fraudster wishes to demonstrate that everything claimed in the letter is true, but once there, the victim could be held, unable to leave the country, until payment has been made for their release. A third possibility is a request made after negotiations have started for the transfer of the monies from Nigeria (in other words, after the victim has swallowed the bait) for the payment of a substantial sum 'to bribe some of the other parties involved'.

Other general points to note are:

- The example letter used here is, it should be said, significantly more literate than most that have come to our notice.
- Postal addresses are often absent, the fraudster preferring telephone and fax numbers for communication.
- Where a postal address *is* given, it only rarely appears in any convincing likeness to a business letterhead. Instead, it gives just the name and address of the individual who is writing – although, considering the nature of the proposal, this is not altogether surprising!
- The letters are usually addressed to *The Chief Executive, The Chairman, The President*, or *The Managing Director*, and only rarely to a named individual – despite the frequent claim that, 'You have been recommended by an associate, who assured me in confidence of your ability and reliability...'

In one memorable version of this scam, the writer stresses how important it is to him 'this time around' to link up with 'a more reliable and trustworthy person' than one, Patrice Miller from New York, with whom last year a similar transaction was carried out. *'With all the documents signed,'* the letter continues, *'the money was duly transferred to his account, only to be disappointed on our arrival in New York and we were reliably informed that Mr Patrice Miller was no longer on that address, while his telephone and telex numbers have been reallocated to somebody else. That is how we lost $27.5 million to Mr Patrice Miller.'*

So, if you should be the lucky recipient of one of these letters, you have been warned!

IMPERSONATION

According to a report in the *Daily Mail* (30 May 1995), a current fraud is based upon stealing the identity of a British citizen in order to obtain goods and services. Having chosen the subject, the fraudster applies to the local post office to have that person's mail redirected to an accommodation address, for which the charge is just £13. The assumed identity is then used to obtain goods from mail order firms, bank and mortgage loans, and credit cards.

Although the criminal intercepts all the mail that is intended for the person whose identity has been assumed, any mail which is not needed or which is irrelevant to the various frauds being perpetrated, the fraudster reseals and posts through the subject's letterbox late at night, so that it seems that it has been delivered with the following morning's mail. By these means, the criminal buys extra time before the impersonation comes to light.

MISREPRESENTATION AND STOLEN CHEQUES

This particular fraud operates through forged UK company letterheads and stolen cheque-books. The example reproduced overleaf is a genuine letter, but did not appear on Kogan Page stationery. It purports to be ordering goods from a Korean manufacturer for despatch direct to its 'Nigerian Branch Office'. As you can see, it requests urgent despatch because of the proximity to the pre-Christmas selling season. In fact, the urgency is required because the fraudsters hope that in this way the goods will have left Korea before the stolen cheque for £2000, sent with the letter, has time to 'bounce'.

KOGAN PAGE
Publishers

PUSAN EEL SKIN(PUSAN GIFT SHOP),
R.C.O. BOX 21, PUSAN,
K O R E A.

Dear Sir,

PROMOTIONAL SALES XMAS TRIAL ORDER, Items: Men's Brief Cases.

Through the Trade Journal of your Country, we understand you are
manufacturers of the above products, as Xmas Sales is fast approaching
we solicite your good offices in selecting quantities good and quality
model and designs for above products to our Sister Company in Nigeria.

Therefore in view of the fast approaching Christmas season, we hereby
enclosed our British Bank Cheque No. 524824 for you to make a very good
selections of the above items for us to enable us equip our stores with
a direct source of products as quoted above at our Nigeria Branch Office.

For this, we leave you with the option of mode of shipment, either by
Airfreight or airmail Express Parcel Post or by Air Courier Services
depending on the one you find cheaper and faster, to be despatched
directly to our Nigeria Branch Office at:

　　　MESSRS: AYO ADEKOYA GENERAL AGENCY, 17, EGBERONGBE, PEDRO ROAD,
　　　　　　　LADY-LAK, SHOMOLU - BARIGA, LAGOS NIGERIA.

Immediately despatch is made in any form, you are to cable the Lagos
Office with the despatch particulars viz; Airway bill/Flight numbers
in case of Airfreight prepaid and date of despatch so for Air courier,
and the Local Agent of the Courier firm in Lagos. Henceforth any more
information or enquiry you may need should be directed to our Sister
Company in Nigeria, at the Telex 20117 NG. Attention: TDS BoX 847.

PAYMENT:- Send the cheque with the Original despatch documents and your
bank particulars to the British Bank under Airmail Registered post
for payment.

DELIVERY:- We want the goods urgently, and after selection and you take
away freight charges, kindly take the balance to pack any quantity that
it could purchase, but with special concession for us and send the goods
to Nigeria as to arrive Nigeria on or before 10th Dec. 1992, before
finalizing Xmas preparation for sales.

Hope to receiving your full co-operation and with best regards.

Yours faithfully,
JOHNSON MATTHEY.

　　　　　　　　　　　　　　　　　enclosed - cheque.

Kogan Page Limited
120 Pentonville Rd London N1 9JN
Tel: 0171-278 0433
Tlx: 263088 KOGAN G Fax: 0171-837 6348

VAT REG NO: 417 8457 28 REG NO 905919 (ENGLAND)

A series of letters have been recovered that have all been sent to Korean firms, all with cheques stolen from the same cheque-book, all valued at £2000, and all requesting delivery to the same address in Nigeria. The goods ordered make fascinating reading and appear to be most suitable for a market-stall trader, namely, men's brief-cases, staplers, wax crayons, 'soccer football size No 5, 32 panels', Sony 5¼ DHSD computer diskettes, and industrial cleaning solvents. In these instances the fraud did not succeed, but how many other letters were sent?

THE FAX DIRECTORY

There are several versions of this fraud, but in essence what is being offered is the inclusion of your company's name and product classification in a 'pan-European'/'British business' fax directory, which may or may not ever be published. Technically, what you receive is a solicitation to advertise in such a publication, but typically this is formatted to look like a trade invoice for a service already provided and is sent direct to the Finance Department. Quite often, either on the first submission or soon after by means of a follow-up 'statement', the document is heavily marked **'OVERDUE'**, threatening 'court action' if payment is not received within a short period. Despite the long-running nature of this scam, it seems that payment is made sufficiently often by busy accounts staff to continue to make it worthwhile to the perpetrators.

The fraud can most readily be combated by ensuring that a proper system for authorising payments exists, which embraces the compulsory matching of an invoice received with an order placed.

Another version of what is essentially the same scam is the solicitation (again disguised as an invoice) to advertise in the **local business directory**, an alternative to the famous *Yellow Pages* directory. Many firms operating in this area do produce a directory of sorts, but its value is something that should be calculated before payment is made.

TELEMARKETING FRAUDS

Typically, these frauds concern office supplies, but they can relate to almost anything which can be made to appear a 'special and unrepeatable offer'. Before the introduction of word processors and laser printers, carbon paper used to be a favourite item for fraudulent sales, the victim usually being the junior employee who looked after stationery supplies. The outcome would be a solicited order and the consequent delivery and (reluctant) payment for a quantity of carbon paper that was sufficient for the next ten years.

Nowadays, speciality advertising and promotional materials – biros, key fobs, calendars, T-shirts and other items overprinted with the corporate name and logo – are the favoured commodities. The solicitation to buy is often accompanied by the promise of a free gift or prize, which usually has a much lower value than that claimed, or is subsequently inflated by unheralded charges that must be met before it can be claimed – eg flight premiums, insurance and hotel upgrades associated with a 'free' holiday. The goods themselves are often of inferior quality and, should the research be undertaken, of higher cost than commonly available elsewhere.

Clearly, corporate rules about purchasing procedures and enforcement of authorisation limits on purchases will largely avoid such problems.

A recent example to come to our attention involved the unsolicited supply of postage meter materials. It began by a telephone call made to the victim company's post-room requesting details of the model number of the franking machine and the name of the person to whom they were speaking. Another call was made the following day, during which they obtained the name of the post-room supervisor. Once these two names were obtained, the fraudster sent supplies, but no invoice, to the post-room. Two or three months later, an invoice was sent for these goods to the firm's accounts department. Needless to say, the prices far exceeded those commonly available in the market-place. At this stage, the invoice was queried because company procedures demanded an official order to support the invoice. The supplier now quoted employees' names, times and dates to support the supply, and it was only the tenacious persistence of the company's security manager which led ultimately to the supplier withdrawing their invoice and reclaiming the unsolicited goods.

MOBILE PHONE FRAUD

When mobile phones first became available they were marketed almost exclusively to the business community. The hardware (the telephone itself) was very expensive, something that clearly limited its adoption by the general public. The investment was, however, generally justified within a business environment by the significant improvement to communications and staff efficiency. Of course, the cost of calls was also expensive compared to fixed line charges, but it was the several hundred pounds cost of a phone that most limited its use in the beginning.

Because of this, the crime most closely associated with mobile phones in the early days was theft of the actual handset, particularly from cars. Mobile phones were extremely desirable, providing significant status for the 'owner'. As far as young street criminals were concerned, possession provided *street cred*, and the possibility of intensive, free short-term use until the service provider was notified of the loss by the original owner and barred further use over the network. When this happened, the criminal just went out and stole another phone.

Within the last year or so, the service providers have changed their marketing strategy to one that seeks to extend mobile phones to a mass market, which inevitably means the general public. To assist this process, mobile phones themselves have been increasingly marketed at minimal cost, each service provider being so keen to achieve high-volume usage on their own network (as opposed to those run by their competitors) that they write off the still high cost of the phones against future rental and call-charge revenues.

There is thus a much-reduced incentive to steal a mobile phone from a car because they can be obtained for virtually nothing from many High Street sales outlets. However, the high costs of network rental and call charges remain, so the criminals' priorities are now directed towards avoiding this aspect of mobile phone usage. Their way of overcoming the problem has been to adopt chip cloning, a technique developed by phone freaks and hackers. Put simply, this operates by copying the electronic code that identifies a particular phone to the network and from whence charges for air-time are calculated. By pretending to be a legitimate user, the criminal obtains free use of his phone – although, of course, the charges incurred find their way to the account of the legitimate user whose identification has been copied.

By limiting the use of each cloned number to a few days, the criminal is very hard to detect. If the cloned number belongs to a business, there is also a good chance that the cost of the criminal's calls will remain unnoticed and will be paid with the rest of that organisation's phone bills. The network managers are aware of this problem, of course, and have their own techniques for detecting unusual call patterns. However, they welcome enquiries from customers whose own suspicions have been raised. The moral, therefore, is to check your telephone bill carefully and report to the network management any unusual and unaccountable charges immediately they are noticed.

BOGUS LOAN BROKERS/ADVANCE FEE FRAUD

There are times when any small and developing business needs access to extra finance. Sometimes this is to fund expansion, sometimes to meet unforeseen cash demands. For the most part, banks are the source of such funding, but there are times when interest rates are high, or when it is the bank itself that has caused the squeeze on cash flow that leads to the need for extra finance, that the entrepreneur will seek alternative and hopefully cheaper sources.

Many will turn to a loan broker as someone who will be able to negoti-
ate alternative finance – and the majority of loan brokers fulfil a legitimate
role. However, there are others, often advertising in local newspapers,
who do not. They will often arrange a meeting in a hotel, rather than in
their office; they will prefer contact by phone, rather than by letter; and
will most probably charge a substantial advance fee before providing any
services.

Legitimate brokers will ask the same kinds of questions as bank managers because they are seeking to place the risk capital available to them
at a rate that will earn their principals a profit, but within an enterprise
where the risk to the investment is considered 'acceptable'. The dubious
broker, on the other hand, will not dwell long on these aspects, preferring
to comment on how positively he views the loan-seeker's business
prospects. He may well also refer to the availability of cheaper foreign
investment finance, may claim that no collateral is required, and will not
be happy at any suggestion that a solicitor or financial adviser should participate in the negotiations.

If there is to be a scam, it will usually occur after the advance fee has
been paid, with the disappearance of the putative loan broker and the
absence of any loan. The 1995 police investigation into the alleged large-
scale frauds perpetrated by and through a German bank operating out of
Paignton in Devon concerns this kind of crime. In this particular case
most of the principals were German nationals, the operation was based
in the UK and most of the victims appear to have been resident in the
USA. The losses are thought to be more than £50 million.

One recent case came to the Middlesex Guildhall Crown Court in
October 1995 and involved the UK company Belling. The alleged perpetrators of the fraud were a British solicitor, another Briton and a US citizen
who, as the following extract from the report of the trial indicates,
obtained a total of £12 million from several victims:

> Belling, the former cooker manufacturer, fell victim to a Europe-wide
> 'advance fee' fraud when its directors were persuaded to pay $3.5 million
> from the company's pension funds in exchange for a future $50 million
> loan, a London court heard yesterday.
>
> Belling was one of a number of companies and individuals duped by fraud-
> sters who claimed high-level contacts with Mr George Bush, the former US
> President, Mr Henry Kissinger, the former US Secretary of State, and the
> British security service MI5, Middlesex Guildhall Crown Court was told.
>
> The victims of the fraud lost a total of £12 million, of which £9 million is still
> missing, the court heard.
>
> Belling directors agreed to pay the $3.5 million fee in May 1991 when the
> company was in financial trouble, the court heard. The loan never materi-
> alised, said Mr John Goldring QC for the Serious Fraud Office. The compa-
> ny never saw its money again and went into receivership the following
> year. (The Financial Times, 1995)

To safeguard your own interests, should you ever seek the help of a loan broker, do obtain the following information **and verify it** before signing any contracts or paying any fees:

░ The name, **address**, and phone number of the broker. (Beware particularly of those who give their address only as a Post Office Box number.)

░ His track record.

░ How much of the broker's income is derived from advance fees? (Insist that any advance payments you make should be placed in an escrow account, that is, a written legal agreement to keep such monies in a third person's account until the agreed service has been provided.)

░ A written statement detailing the arrangements that will apply over the refund (or not) of advance fees, if no loan is eventually brokered.

░ A written prospectus describing the broker's full range of services.

PRODUCT DIVERSION

This fraud depends upon the industry practice that sets different prices for the same product as a way of breaking into new markets, or countering intense competition that might develop in a foreign market and keeps prices down. There are often tax incentives and other cost savings that allow manufacturers to sell their products to overseas distributors at significantly lower prices than those obtainable by distributors in the manufacturer's home country. Product diversion operates by illegally rerouteing goods that are intended for export *via* a third party back into the country of origin. The cheaper prices that apply then enable these goods (which are often high-value or luxury items, such as perfumes and branded clothing) to reach the retail market at prices which significantly undercut the established, legitimately obtained products. The manufacturer loses through depressed sales of his products at the full established price within the home market.

The warning signs that this might be happening to you include:

░ A buyer insists that the purchased goods be shipped air-freight, when the quantities involved would normally warrant sea-/surface-freight, to save costs.

░ Sales levels in foreign countries which are at unexpectedly, even illogically, high levels.

░ A purchaser's insistence that the product labelling is in English, when that is not the language used in the destination country.

░ Sales which are intended for distribution to countries with bad relations with the manufacturing country, eg Britain and Iraq, France and Australia/New Zealand.

- Luxury items which are purchased for sale in impoverished countries.
- Sales which are sent to countries at war.

SHORT AND TIME-EXPIRED DELIVERIES

It was Gerald Mars in his book *Cheats at Work* who identified what he termed *triadic* occupations, where there are three-way relationships (employer–employee–customer) which enable alliances to be made by any two parties against the third. Such occupations are a feature of the personal service industries, where the most common fiddles occur through collusion between employers and employees, who are able to steal from customers by short-changing them or by making short deliveries.

Sometimes, as in date-stamped items, such as perishable foods, where there is a sale-or-return agreement between the customer and the supplier, the 'fiddle' involves recycling already time-expired goods that have been picked up from customer A by delivering them to customer B. Where such practice is endemic, the drivers have the equipment to remove the original date from the packaging and restamp it with a new one. One driver is alleged to have commented that nobody worried about this unless the bacon actually turned green!

Mars quotes from Ditton's treatise, *Part-Time Crime*, where it was pointed out that a baker's roundsman on regular deliveries (particularly where he sells to retailers) is forced, in effect, into fiddling because he is made accountable for shortages on his round.

There are always shortages because the system of accounting is inevitably geared against him. In addition there is the likelihood that he can be short after being fiddled by the loaders of his vehicle or by his customers. Ditton carefully describes the collusive training whereby managements teach their staffs how to overcome the liabilities by learning to fiddle customers. (Mars, 1994, pp 149–150)

Clearly, the potential for this kind of crime extends over a wide variety of businesses, and illustrates the importance of being aware of the vulnerabilities of the situation, one of which is the familiarity that derives from the regular attendance, day after day, of the same delivery person on your premises, and having in place an effective system for receiving goods deliveries.

References

Financial Times, 'Advance fee fraud is alleged: Belling duped out of $3.5m, court hears', court report in the *Financial Times*, Wednesday 4 October 1995, London.

Mars, Gerald (1994)   *Cheats at Work* (2nd edn), Dartmouth Press, Aldershot.

13

USEFUL CONTACTS AND ADDRESSES

Access Control

National Approval Council for Security Systems (NACOSS)
Queensgate House
14 Cookham Road
Maidenhead
SL6 8AJ
Tel: 01628 637512
Fax: 01628 773367

Security Systems and Alarms Inspection Board
6 Northumberland Place
North Shields
Tyne and Wear NE30 1QP
Tel and fax: 0191 296 3242

Agencies and Official Bodies

Audit Commission
1 Vincent Square
London SW1P 2PN
Tel: 0171 828 1212

British Standards Institution
389 Chiswick High Street
London W4 4AL
Tel: 0181 996 9000
Fax: 0181 996 7400

Crime Concern
Signal Point
Station Road
Swindon SN11 1FE
Tel: 01793 514596

Data Protection Registrar
Wycliff House
Water Lane
Wilmslow SK9 5AF
Tel: 01625 545745

Health and Safety Commission
Baynards House
1 Chepstow Place
London W2 4TF
Tel: 0171 243 6000

Joint Security Industry Council
Box 17324
London EC1A 7NA
Tel: 0800 731 4571
Fax: 01628 773367

Loss Prevention Council
Melrose Avenue
Borehamwood WD6 2BJ
Tel: 0181 207 2345

National Audit Office
157–197 Buckingham Palace Road
London SW1W 9SP
Tel: 0171 798 7000

Risk and Security Management Forum (RSMF)
c/o 53 Elmcroft Drive
Chessington
KT19 1DY
Tel: 0181 391 9290

Alarm Receiving Centres

National Approval Council for Security Systems (NACOSS)
Queensgate House
14 Cookham Road
Maidenhead
SL6 8AJ
Tel: 01628 637512
Fax: 01628 773367

Alarms
(see Intruder Alarms)

CCTV

National Approval Council for Security Systems (NACOSS)
Queensgate House
14 Cookham Road
Maidenhead
SL6 8AJ
Tel: 01628 637512
Fax: 01628 773367

Security Systems and Alarms Inspection Board
6 Northumberland Place
North Shields
Tyne and Wear NE30 1QP
Tel and fax: 0191 296 3242

Central Monitoring Stations (see Alarm Receiving Centres)

Certification Body (for Installers of Security Technology)

National Approval Council for Security Systems (NACOSS)
Queensgate House
14 Cookham Road
Maidenhead
SL6 8AJ
Tel: 01628 637512
Fax: 01628 773367

Closed Circuit Television (see CCTV)

Computers

National Computing Centre
Oxford Road
Manchester
M1 7ED
Tel: 0161 228 6333

Contract Guarding (see Guarding Services)

Trade and Industry, Department of Information Technology Division
151 Buckingham Palace Road
London SW1W 9SS
Tel: 0171 215 5000

Document Destruction

Data Disposal Ltd
Osier Way
London E10 5SB
Tel: 0181 556 5608
Fax: 0181 558 9230

Education and Training

Correspondence Training College
IPSA House
3 Dendy Road
Paignton
TQ4 5DB
Tel: 01803 5548549

Electrical Installation Engineering Industry Training Organisation (EIEITO)
ESCA House
34 Palace Court
London W2 4HY
Tel: 0171 229 1266
Fax: 0171 221 7344

Fire Protection Association
Melrose Avenue
Borehamwood
WD6 2BJ
Tel: 0181 207 2345

Group 4 Securitas Training Ltd
Farncombe House
Broadway
Worcestershire WR12 7LJ
Tel: 01386 858585

International Professional Security Organisation
3 Dendy Road
Paignton
Devon TQ4 5DB
Tel: 01803 554849

Leicester University
Centre for the Study of Public Order
154 Upper New Walk
Leicester LE1 7QA
Tel: 01162 522458

Loughborough University of Technology
Centre for Hazard and Risk Management
Loughborough LE11 3TU
Tel: 01509 222151

Securiguard Services Ltd
Templar House
1 Boycott Avenue
Oldbrook
Milton Keynes MK6 2RW
Tel: 01908 668576

Security Industry Training Organisation
Security House
Barbourne Road
Worcester WR1 1RT
Tel: 01905 20004

Electronic Surveillance Countermeasures

Audiotel International Ltd
Cavendish Courtyard
Sallow Road
Weldon Industrial Estate
Corby NN17 5DZ
Tel: 01536 266677

Fire

Fire Protection Association
Melrose Avenue
Borehamwood
Herts WD6 2BJ
Tel: 0181 207 2345

Fire Research Station
Borehamwood
Herts WD6 2BL
Tel: 0181 953 6177

Institution of Fire Engineers
148 New Walk
Leicester LE1 7QB
Tel: 0116 255 3654

Government

Companies House
Crown Way
Cardiff CF4 3UJ
Tel: 01222 388588

Home Office
50 Queen Anne's Gate
London SW1H 9AT
Tel: 0171 273 3000

Office of Fair Trading
Field House
15–25 Bream's Buildings
London EC4A 1PR
Tel: 0171 242 2858

Registry of County Court Judgements
173–175 Cleveland Street
London W1P 5PE
Tel: 0171 380 0133

Trade and Industry, Dept of, Investigation Division
1 Victoria Street
London SW1H 0ET
Tel: 0171 215 5000

Trade and Industry, Dept of, Information Technology Division
151 Buckingham Palace Road
London SW1W 9SS
Tel: 0171 215 5000

Guarding Services

British Security Industry Association Ltd
Security House
Barbourne Road
Worcester WR1 1RT
Tel: 01905 21464

International Professional Security Organisation
3 Dendy Road
Paignton
Devon TQ4 5DB
Tel: 01803 554849

Investigations

Association of British Investigators
ABI House
10 Bonner Hill Road
Kingston upon Thames
KT1 3EP
Tel: 01482 224488
Fax: 01482 218715

Control Risks Group
83 Victoria Street
London SW1H 0HW
Tel: 0171 222 1552
Fax: 01803 529203

Institute of Professional Investigators
31A Wellington Street
St John's
Blackburn
BB1 8AF
Tel: 01254 680072
Fax: 01254 59276

Network International
26 Dover Street
London W1X 4JU
Tel: 0171 344 8100
Fax: 0171 344 8101

Intruder Alarms

National Approval Council for Security Systems (NACOSS)
Queensgate House
14 Cookham Road
Maidenhead
SL6 8AJ
Tel: 01628 637512
Fax: 01628 773367

Security Systems and Alarms Inspection Board (SSAIB)
6 Northumberland Place
North Shields
Tyne and Wear NE30 1QP
Tel and fax: 0191 296 3242

Journals (see Media)

Locks and Safes

Master Locksmiths Association
Units 4 & 5
The Business Park
Woodforde Halse
Daventry
NN11 3PZ
Tel: 01327 262255

Media

Computer Audit Update
Elsevier Advanced Technology
PO Box 150
Kidlington
Oxford OX5 1AS
Tel: 01865 843000

Computers and Security
Elsevier Advanced Technology
PO Box 150
Kidlington
Oxford OX5 1AS
Tel: 01865 843848

Fire Prevention
Official Journal of the Fire Protection Association
Loss Prevention Council
Melrose Avenue
Borehamwood WD6 2BJ
Tel: 0181 207 2345

International Security Review
FMJ International Publications
Queensway House
2 Queensway
Redhill RH1 1OS
Tel: 01737 768611

Intersec
Three Bridges Publishing Ltd
Bridge House
Aviary Road
Pyrfod
Woking GU22 8TH
Tel: 01932 340418

Police Review
South Quay Plaza 2
183 Marsh Wall
London E14 9FZ
Tel: 0171 537 2575

Professional Security
JTC Associates Ltd
4 Elms Lane
Shareshill
Wolverhampton WV10 7JS
Tel: 01922 415233

Security Management Today
Blenheim House
630 Chiswick High Road
London W4 5BG
Tel: 0181 742 2828

Police

Association of Chief Police Officers
(England, Wales and Northern Ireland)
Wellington House
63 Buckingham Gate
London SW1E 6BE
Tel: 0171 230 7184

Association of Chief Police Officers
(Scotland)
Police Headquarters
Fettes Avenue
Edinburgh EH4 1RB
Tel: 0131 311 3131

Home Office Crime Prevention College
Harkhills
Easingwold
York YO6 3EG
Tel: 01347 825060
Fax: 01347 825099

Home Office Police Department (F8)
Queen Anne's Gate
London SW1H 9AT
Tel: 0171 273 3000

National Criminal Intelligence Service
PO Box 8000
Spring Gardens
Tinworth Street
London SE11 5EN
Tel: 0171 238 8000

Police Research Group
Home Office
50 Queen Anne's Gate
London SW1H 9AT
Tel: 0171 273 3000

Police Scientific Development Branch
Home Office
Woodcock Hill
Sandridge
St Albans AL4 9HQ
Tel: 01727 865051

Professional Associations

American Society for Industrial Security (ASIS)
UK Chapter 208
Vice Chairman and Secretary
Geoff Whitfield
Company Security Manager
Glaxo Wellcome Research &
Development
Gunnels Wood Road
Stevenage SG1 2NY
Tel: 01438 745745

Association of British Insurers
51 Gresham Street
London EC2V 7HQ
Tel: 0171 600 3333
Fax: 0171 696 8999

Association of British Investigators
ABI House
10 Bonner Hill Road
Kingston upon Thames
KT1 3EP
Tel: 01482 224488
Fax: 01482 218715

Association of Consulting Engineers
Alliance House
12 Caxton Street
London SW1H 0QL
Tel: 0171 222 6557

Association of Security Consultants
The Secretary
Rushton Hall
Tarrant Rushton
Blandford Forum
DT11 8SD
Tel: 01258 450066
Fax: 01258 480168

British Computer Society
Station Road
Swindon SN1 1EZ
Tel: 01793 480269

EPIC (Ex-Police in Industry and Commerce)
5 Melrose Road
Bishop Monkton
Harrogate
HG3 3RH
Tel and fax: 01765 677408

Guild of Security Controllers
c/o GEC Marconi Sonar Systems
Division Ltd
Wilkinthroop House
Templecoombe
BA8 0DH
Tel: 01935 442027
Fax: 01935 442317

Institute of Insurance Consultants
PO Box 381
121a Queensway
Bletchley
Milton Keynes MK1 1XZ
Tel: 01908 643364

Institute of Internal Auditors – United Kingdom
13 Abbeville Mews
88 Clapham Park Road
London SW4 7BX
Tel: 0171 498 0101
Fax: 0171 978 2492

Institute of Hotel Security Management
London Marriott Hotel
Grosvenor Square
London W1A 4AW
Tel: 0171 493 1232
Fax: 0171 491 3201

Institute of Professional Investigators
31A Wellington Street
St John's
Blackburn
BB1 8AF
Tel: 01254 680072
Fax: 01254 59276

Institute of Security Management
17 Hough Road
Kings Heath
Birmingham
B14 6HL

International Professional Security Association (IPSA)
3 Dendy Road
Paignton
Devon TQ4 5DB
Tel: 01803 554849

Master Locksmiths Association
Units 4/5 The Business Park
Woodford Halse
Daventry
Northants NN11 3PZ
Tel: 01327 262255

Trade Associations

Association of British Insurers
51 Gresham Street
London EC2V 7HQ
Tel: 0171 600 3333
Fax: 0171 696 8999

British Retail Consortium
Bedford House
69/79 Fulham High Street
London SW6 3JW
Tel: 0171 371 5185

British Security Industry Association Ltd
Security House
Barbourne Road
Worcester WR1 1RT
Tel: 01905 21464

Electrical Contractors Association
34 Palace Court
London W2 4HY
Tel: 0171 229 1266
Fax: 0171 221 7344

Fire Protection Association
Melrose Avenue
Borehamwood WD6 2BJ
Tel: 0181 207 2345

Training (see Education and Training)

Vehicle Security

Maple & Son Ltd
Commercial Vehicle Security
Specialists
Unit 3, Sheffield Street
Heaton Norris,
Stockport SK4 1RU
Tel: 0161 477 3476

Index

References in italics indicate figures or tables.

Index of Advertisers

Visit Kogan Page on-line

Comprehensive information on
Kogan Page titles

Features include

- complete catalogue listings,
  including book reviews and
  descriptions

- special monthly promotions

- information on NEW titles and
  BESTSELLING titles

- a secure shopping basket facility
  for on-line ordering

PLUS everything you need to know
about KOGAN PAGE

http://www.kogan-page.co.uk